Charles James Cullingworth

The Values of Abdominal Section in Certain Cases of Pelvic

Peritonitis,

Based on a Personal Experience of Fifty Cases

Charles James Cullingworth

The Values of Abdominal Section in Certain Cases of Pelvic Peritonitis, *Based on a Personal Experience of Fifty Cases*

ISBN/EAN: 9783744743730

Printed in Europe, USA, Canada, Australia, Japan

Cover: Foto ©berggeist007 / pixelio.de

More available books at **www.hansebooks.com**

THE

VALUE OF ABDOMINAL SECTION

IN

CERTAIN CASES OF PELVIC PERITONITIS,

BASED ON A PERSONAL EXPERIENCE OF FIFTY CASES.

BY

CHARLES J. CULLINGWORTH, M.D., F.R.C.P.

Read October 5th, 1892.

[*Reprinted, for private circulation, from Volume XXXIV of the 'Transactions of the Obstetrical Society of London.'*]

LONDON:
PRINTED BY ADLARD AND SON,
BARTHOLOMEW CLOSE, E.C., AND 20, HANOVER SQUARE, W.

1893.

THE VALUE OF ABDOMINAL SECTION IN CERTAIN CASES OF PELVIC PERITONITIS, BASED ON A PERSONAL EXPERIENCE OF FIFTY CASES.

By CHARLES J. CULLINGWORTH, M.D., F.R.C.P.

(Received Sept. 19th, 1891, and Feb. 20th, 1892.)

(Abstract.)

THE question considered in this paper is whether surgical interference is or is not frequently called for in cases of pelvic peritonitis. The author answers this question in the affirmative, and supports his opinion by a detailed record of fifty cases in which he has himself operated. The paper is accompanied with a table, showing for each case the symptoms, the physical signs, the diagnosis, the actual condition disclosed at the operation, the nature of the operation performed, and the results, immediate and (where possible) remote. The cases are arranged in the order of their occurrence, their classification being reserved for the concluding part of the paper. This method seems to be the best suited for showing the gradual development of the author's present views and practice, and at the same time serves to emphasise the fact that a correct classification can only be made after the diagnosis has been tested by actual inspection of the diseased parts.

The cases include the whole of the author's experience of the operation up to the end of February, 1891, and are classified as follows:

Suppurating salpingitis 20
Non-suppurating salpingitis, including six cases complicated with
    suppurating ovarian cyst 12
Tubercular disease of Fallopian tubes 2

Pelvic abscess, seat undetermined 3
Pedunculated retro-peritoneal cyst, with abscesses in walls . . 1
Tubercular abscess in abdominal wall, with masses in pelvis (tubercular glands) and miliary tubercle of peritoneum . . . 1
Hæmatocele 2
Hæmatosalpinx with hæmatocele 3
Hæmatoma of broad ligament 1
Broad ligament cysts:
  (a) With ovaritis 2 ⎱
  (b) With hydrosalpinx 1 ⎰ 3
Encysted peritonitic effusion 1
Retroflexed uterus with fibroids 1
                                                       —
                                                     50

Pelvic peritonitis was common to all the cases except the last-named (Case 32), in which an erroneous diagnosis was made.

The cases of suppurating salpingitis are subdivided as follows:

(a) With occlusion (pyosalpinx) (Cases 7, 15, 30, 40, 43) . . 5
(b) With distal end open (Cases 16 and 36) . . . 2
(c) With suppurative disease of the ovary (Case 37) . . 1
(d) With a direct communication between the tube and a suppurating cyst of the adjacent ovary (suppurating tubo-ovarian cyst) (Cases 17, 18, 20, 25, 33, 50) 6
(e) With non-suppurating cystic ovary (Case 27) . . . 1
(f) With suppurating hæmatocele (Case 14) . . . 1
(g) With hydrosalpinx (Cases 9 and 45) 2
(h) With intra-peritoneal abscess (Cases 28 and 49) . . 2
                                                     20

The cases of non-suppurating salpingitis are subdivided as follows:

(a) Uncomplicated cases (Cases 19 and 24) 2
(b) With suppurating ovarian cyst (Cases 4, 12, 26, 39, 41, 48) . 6
(c) With non-suppurating ovarian cyst (Cases 35 and 46) . . 2
(d) With hæmatosalpinx and hæmorrhagic ovarian cyst (Case 2) . 1
(e) With double hæmatocele (Case 11) 1
                                                     12

Pelvic suppuration was present in thirty cases, or 60 per cent. It occurred in the Fallopian tube alone in thirteen cases, in the ovary alone in six cases, in both tube and ovary in seven cases (in six of which tube and ovary were in direct communication), while in the remaining four cases the seat of suppuration was either not precisely determined or did not involve either the tube or the ovary.

There was strong presumptive evidence of gonorrhœa in a large proportion of the cases, and in at least five cases the proof seemed complete.

Nine of the cases died, a mortality of 18 per cent. Seven of the deaths were due to peritonitis, probably septic, one to acute nephritis, and one to collapse on the eleventh day.

Of the fatal cases one was tubercular disease of the tubes, two were purulent salpingitis, one was double salpingitis with old hœmorrhage, two were suppurating tubo-ovarian cysts, one was retro-peritoneal suppurating cyst, two were old peritonitis with serous cysts of broad ligament.

As experience increased, the mortality became sensibly diminished.

Hæmorrhage, to a greater or less extent, existed in twelve of the thirty-two cases of salpingitis. In five cases there was amenorrhœa, in three dysmenorrhœa, whilst in twelve the menstrual function was undisturbed.

In sixteen cases the removal of the appendages was complete, in twenty-three partial.* Of the former, fifteen recovered; of the latter, seventeen.

The peritoneum was flushed in twenty-two cases, of which eighteen recovered.

Drainage was employed in forty-seven out of the fifty cases.

In two cases a fæcal fistula formed, which in each instance healed spontaneously.

In five cases the patients complained some time after the operation of more or less persistent pain.

A sinus existed in two of the cases when the patients were last seen.

In four cases a hernia has occurred in the line of incision.

Attention is called to the unreliability of the temperature as a sign of the existence of pelvic suppuration, the temperature before operation having been absolutely normal in twelve of the thirty cases in which suppuration was present.

* By "complete" is here meant bilateral, and by "partial" unilateral.

In the course of the remarks appended to each case the following incidental propositions are laid down, either directly or by inference :

1. Recurrent attacks of pelvic peritonitis in the female ought always to lead to a strong suspicion of the existence of chronic disease of the uterine appendages, and to careful bimanual examination.

2. Purulent collections in the pelvis are particularly apt to set up recurrent peritonitis, and are more common than is usually supposed.

3. Where distinct swellings are found in the posterior quarters of the pelvis, in connection with recurrent attacks of pelvic peritonitis, surgical relief is usually indicated, and, generally speaking, the sooner such relief is afforded the better.

4. Purulent inflammation of the mucous membrane of the Fallopian tube differs from purulent inflammation of other mucous membranes in the absence, owing to the anatomical situation of the Fallopian tubes, of a natural outlet for the pus. A very slight amount of swelling of the mucous membrane suffices to block the tube at its uterine end, and if pus be present in the tube, it must then either remain pent up in the tube, or be poured out through the fimbriated end into the peritoneum, in either case becoming a source of danger.

5. Salpingitis being a painless affection, the wall of a pyosalpinx may be on the point of perforation before an acute attack of peritonitis gives warning of the presence of serious disease.

6. It is safer to attack cases of pelvic suppuration from above than from below.

7. Suppurating tubo-ovarian cysts are usually the result of ulceration on the tubal side of the adhesion between tube and ovary, but in exceptional cases result from ulceration on the ovarian side.

8. The immediate results are more satisfactory after complete (bilateral) than after partial (unilateral) operations.

9. One of the chief risks in the operation for the separation and removal of inflamed tubes is the liability to mistake thickened and adherent intestine for diseased tube. The way to avoid error is to trace the tube from its uterine end outwards.

10. The exceptional instances in which pain persists after operation for gross lesions of the uterine appendages are generally to be explained either by omental or intestinal adhesions, or by the co-existence with the actual disease of a neurotic condition, of which the pelvic pain is a mere local expression.

11. Tubal disease in the virgin is generally, if not always, tubercular.

12. Hydrosalpinx, in the great majority of cases, is merely a form of retention-cyst, due to occlusion of the distal end of the tube from without.

13. Simple collections of serum, both large and small, are apt to form beneath the peritoneum covering the tube and broad ligament in chronic cases of pelvic inflammation, especially in those of very long standing. Probably the best treatment of these cysts, after exposing them and making certain of the diagnosis by abdominal section, is simple puncture and evacuation, the risk of removal being, in the author's experience, out of proportion to their importance.

14. Hæmatosalpinx, though no doubt due, in the majority of cases, to tubal gestation with apoplexy of the ovum, is sometimes an incident in the course of a chronic salpingitis. In these exceptional cases the walls of the distended tube, instead of being attenuated by the distension, as Bland Sutton has shown them to be in tubal gestation, are thickened by inflammatory deposits.

Part I.—Cases 1 to 25.

This paper is offered as a contribution towards the settlement of a question that has been for several years hotly debated, both in this country and in America, namely, whether surgical interference is or is not frequently called for in cases of pelvic inflammation.

The discussion has, in this country, recently assumed a phase that makes it incumbent on those of us who have any evidence to bring forward to do so with as little delay as possible. I propose, in this communication, to approach the subject solely from the point of view of my own experience, an experience that, I venture to think, is now sufficiently extensive to justify me in laying my results before the Society.

It has been a matter of much difficulty to decide in what order the cases should be arranged. All things considered, it has appeared to me best to present them in the order of their occurrence. By this plan the Society will be enabled to follow the steps by which I have been gradually led to the adoption of my present views, and to judge how far those views are warranted by the teachings of my own experience. Moreover any attempt at classification must necessarily be based upon knowledge obtained during the operation, and would therefore fail to convey a correct impression of the difficulty of the problem that confronts us at the bedside and in the consulting-room. If these cases could all be accurately diagnosed and classified before operation, our task would be much easier.

But although great advances have recently been made

in the diagnosis of intra-pelvic disease, the most experienced amongst us will acknowledge that it is not yet possible to make out the precise condition of the parts in every case of pelvic inflammation. We cannot even always distinguish with certainty between purulent and non-purulent cases. If we could, the scope of the discussion would be much narrower. Indeed, I am inclined to think that we should then all agree. In the meantime we must take things as they are, and, recognising our deficiences both in knowledge and in power of observation, make allowance, in any rules we may lay down, for occasional errors of diagnosis.

I am sorry to have to burden my paper with the details of so many cases. But without details the communication, regarded as a piece of evidence, would be worthless. In the accompanying table are presented the main points in each case, viz. the circumstances that induced me to operate, the nature of the operation, the actual condition found, and the result.

Several of the earlier cases have already appeared in print. The inclusion of these in the tables requires no apology, but the fact that some of them are again related with full details in the paper itself seems to call for a few words of explanation. I should have been glad, both for the sake of shortening my paper and avoiding repetition, to omit them; but the object of this communication being to present a complete and faithful history of my personal experience, it seemed to me better to tell the story of some of my cases over again, than, by omitting them, to mar the completeness and so lessen the value of the record.

With the exception, therefore, of five cases (of which the particulars have been published quite recently, and the references to which are given in the table), this paper includes, in more or less detail (sufficient, I hope, for purposes of criticism and discussion), an account of every case in which I performed abdominal section for the relief of pelvic inflammation up to the end of February, 1891.

I had been operating for nearly twelve years, in cases

of ovarian and other abdominal tumours, before I ventured to open the abdomen in a case of intra-pelvic disease where there was no abdominal tumour. There had been for some time a growing conviction in my mind that such operations ought to be undertaken; but, being somewhat slow to take up new methods of treatment, it was several years before conviction ripened into action. At length a typical case presented itself.

CASE 1.* *Symptoms of pelvic peritonitis for six years; swelling on both sides of the uterus, more marked on right; abdominal section; chronic ovaritis on right with polycystic tumour of each broad ligament; removal of tumours and of right tube and ovary; recovery; pain entirely relieved.* —Annie McC—, aged 25, applied at the out-patient department of St. Mary's Hospital, Manchester, on account of constant pain and sensation of weight in the lower part of the abdomen, rendering her quite unable to continue her calling as a dressmaker. She was married at the age of seventeen, had never been pregnant, and had now been a widow for three years. The pain commenced six years ago, and had continued ever since with one or two short intervals; it was most severe on the left side. She had consulted several eminent gynæcologists in London, and had at one time been a patient at the Chelsea Hospital, where she obtained considerable temporary relief. But the symptoms returned when she resumed her ordinary life, and increased in severity from year to year until, twelve months ago, she found she was unable to maintain the sitting posture sufficiently long to continue her occupation. During the last six months she had earned what she could as an artist's model. She had an anæmic and careworn appearance, and her general health was evidently becoming impaired.

* An account of this and the following case was published in a paper entitled " Abdominal Section for the Removal of Small Intra-pelvic Tumours of the Ovaries and Adjacent Parts, with Notes of Two Cases," ' Brit. Med. Journ.,' January 30th, 1886.

On bimanual examination of the pelvis, a firm, rounded, tender swelling was felt to the right of and slightly behind the uterus; the uterus itself was normal in size and position. The patient attended the outdoor department for about seven weeks, and, as she did not in any way improve, I suggested an exploratory incision, with a view to removing the disease, if it were found practicable. As her life was a burden to her, and she was unfit for any kind of work, she readily consented to run the risk of the operation; and accordingly I admitted her as an in-patient on May 11th, 1885, and explored the abdomen with antiseptic precautions on the 13th.

I expected to find a chronically inflamed and enlarged ovary on the right side, and an inflamed and adherent ovary without marked enlargement on the left. What I did find was as follows: on the right side a chronically inflamed and adherent ovary of the size of a walnut, and in addition to this a firm tumour of the broad ligament, of the size of a closed fist, consisting of a compact mass of exceedingly small cysts; on the left side another broad ligament tumour, of similar character to that on the right side, but smaller. The left ovary was apparently healthy. I enucleated both the broad ligament tumours, and removed the right ovary with part of the Fallopian tube, leaving the left ovary and tube undisturbed. The operation was rendered somewhat difficult by numerous very firm adhesions. A glass drainage-tube was inserted and left in for forty-eight hours. The temperature rose to 102° F. in the evening of the day of operation, but soon fell to 100° F.; and although it rose on the morning of the fifth day, and again on the morning of the sixth day, to 101° F., it did not again occasion the least anxiety, and the patient made an excellent recovery.

I saw her seven months later. Her only complaint then was that she menstruated too frequently. She had lost her anæmic appearance, and had become stout and well, and being entirely relieved of her pain, she was now able to follow in comfort her occupation as a dressmaker.

It will be observed that, in this case, two small tumours were found, one in each broad ligament. But as these were not diagnosed, and the operation was performed under the impression that the whole of the mischief was of inflammatory origin, the case is evidently entitled to a place in this series. No mention is made of the condition of the tubes. I was not at that time alive to the importance of tubal inflammation as a precursor of pelvic peritonitis. As often happens, the pain was on the opposite side to that on which the disease was most marked. This is a clinical fact that I am unable to explain. I am content to know that the pain disappeared when the disease was removed.

CASE 2. *Severe dysmenorrhœa for seven years; continuous pain with hæmorrhage for two months; tender, firm, oblong swelling on right side displacing uterus to left; abdominal section; blood-cyst of right ovary, smaller cyst of left; chronic inflammation of right tube, with hæmatosalpinx, left tube healthy; both ovaries and right tube removed; recovery.*—Mary M—, aged 26, married to a winder in a cotton mill, was admitted into St. Mary's Hospital, Manchester, on September 25th, 1885, complaining of continuous pain in the lower part of the abdomen, especially on the right side and down the right thigh. The pain had existed for seven years, commencing soon after the birth of her only child. At first it only came on immediately before each menstrual period, but even then it was so severe while it lasted that she was rendered unfit for work. During the last two months the pain had been severe and continuous, and there had been persistent hæmorrhage from the uterus.

The patient on admission was thin and anæmic, with a haggard and pinched countenance, betokening much suffering. On bimanual examination of the pelvis the right side was found to be occupied by an oblong, firm swelling, very tender to the touch, pushing over the uterus to the left of the middle line. The diagnosis was uncertain, but I thought it most probable that there was distension of the right Fallopian tube. The hot douche and absolute rest were found, at the end of a fortnight, not to have resulted in the least relief; and accordingly, the risk having been explained to the patient, an exploratory incision was made in the middle line of the abdomen on October 7th. The right ovary was found to be enlarged to the size of a hen's egg, and to be cystic; the contents of the cyst, which escaped during removal, consisted of dark fluid blood altered by long retention. Closely connected with the diseased ovary was a thick fusiform swelling, consisting of the Fallopian tube distended with blood, partly fluid and partly clotted, the walls of the tube being much thickened by chronic inflammation, and firmly adherent externally to a coil of small intestine. After carefully separating the adhesions the tube and ovary were both removed, the ligature being placed close to the uterus. The left ovary was also found to be enlarged from incipient cystic disease, and was accordingly removed. The tube on the left side was healthy. A glass drainage-tube was inserted at the lower angle of the wound, and was allowed to remain until the fourth day. The patient made an excellent recovery, the temperature only once rising to 100° F. She had some pain about a fortnight after the operation, but it soon passed off, and in the month of December she had become entirely free from pelvic discomfort, and was able to go about as usual.

This was a case of chronic unilateral salpingitis, in the course of which hæmorrhage had occurred, distending the tube with blood. Such cases are distinguished from hæmatosalpinx due to apoplexy of the ovum in a tubal gestation, not only by the discovery of chorionic villi in the latter, but also by the condition of the walls of the tube, which in cases of hæmorrhage due to tubal gestation, are, as Bland Sutton has pointed out,* abnor-

* "It is a fact important to be remembered that when a Fallopian tube becomes distended by fluid accumulations, or even by an impregnated ovum

mally thin instead of being abnormally thick. In the one there is simple distension with, at the most, some turgescence; in the other there is inflammation as well as distension. The co-existence, in cases of inflammatory hæmatosalpinx, of blood-cysts in the adjacent ovary is by no means infrequent. Several additional examples will be given in the course of this paper.

CASE 3. *Recurrent pelvic peritonitis for ten years; constant pain in left iliac region and back, with discharge of blood from rectum and pain on defæcation, for five years; thickening in situation of both broad ligaments; prolapsed and adherent left ovary; abdominal section; chronic pelvic peritonitis, ovaries normal, adherent; left broad ligament thickened, right tube distended with serum, three cysts in right broad ligament; cysts and right tube removed; death; autopsy.*—J. R—, aged 35, married, housekeeper, was admitted into St. Mary's Hospital, Manchester, January 12th, 1886, on account of severe pain in left iliac region. She had been married eighteen years, and had borne two children, the last one fifteen years ago. Her health had been exceedingly good up to ten years ago, when she had an attack of peritonitis, and was confined to bed altogether for about five months. She had a considerable quantity of vaginal discharge and also a good deal of bleeding and purulent discharge, from the bowel. A year or two later she began to suffer severe pain in the left iliac region. At first this only came on immediately before each menstrual period; after a short time it became constant, though it was always worse at the periods. Five years ago she was again laid up for a considerable time. On leaving the hospital she became an out-patient, and she has attended more or less regularly ever since. The pain has gradually become more

developing within it, the walls of the tube gradually thin. In this respect the tubes are in striking contrast with the uterus."—" Lecture on the Value of Comparative Pathology to Philosophical Surgery," 'Brit. Med. Journ.,' February 21st, 1891, p. 398.

severe and constant, and is felt in the back as well as the iliac region. The patient has been entirely unable to undertake ordinary housework for several years, and her suffering is often exceedingly severe. Lately she has lost flesh. Menstruation is, for the most part, regular; during the last month there has been some irregular hæmorrhage.

On admission there is nothing abnormal to be detected on examination of the abdomen.

Per vaginam, os uteri patulous, old laceration of cervix on left side. Uterus retroverted and slightly displaced to right; swelling in Douglas's pouch consists of corpus uteri. The left broad ligament gives the sensation of being thickened, and a small body, tender to the touch, is felt behind it, close to the uterus. There is very slight thickening in the region of the right broad ligament; a soft cord can be felt, like the Fallopian tube. The diagnosis was chronic ovaritis of left side, with extensive adhesions. The abdomen was opened on the 13th of January. The contents of the pelvis were much matted. The uterus was retroverted and fixed by adhesions. There was no cyst or tumour detected on the left side. Both ovaries appeared to be normal. In the right broad ligament three cysts were found of varying size, the largest being about equal in size to a goose's egg. The smallest cyst appeared to be in direct communication with the interior of the Fallopian tube, which was distended with serum. The parts removed consisted of the tube and broad ligament cysts from the right side. A drainage-tube was inserted, and the wound closed. There was a good deal of pain and a little sickness during the first forty-eight hours, but it was not until the morning of the fourth day that the patient's condition gave rise to serious anxiety. The temperature, hitherto under 100°, gradually rose, the pulse became rapid, and there was constant retching. She died a little after midday.

On post-mortem examination the following day the omentum was found thickened and hyperæmic. A band

passed down from it into the left side of the pelvis, where
it was firmly adherent. There were two or three fluid
ounces of blood-stained serum in the peritoneal cavity,
but there was no evidence of suppuration there or elsewhere. The pouch of Douglas was obliterated by the
retroverted and adherent uterus. On the right side
there was a large adherent blood-clot just above the
ligature; no ovary could be found on that side. On the
left side there was considerable thickening of the broad
ligament; the left ovary was slightly enlarged. The
intestines were considerably distended, their serous coat
showing signs of commencing inflammation. There was
an abrasion of the outer coat of the ileum, about a quarter
of an inch in diameter, situated about five or six inches
from the cæcum. Old adhesions existed between the
coils of intestine in the upper part of the abdomen and
between intestine and omentum. The intestinal canal
was opened from pylorus to rectum, no stricture or ulcer
being discovered. The liver, kidneys, spleen, pancreas,
and stomach showed no morbid change.

In this case I was surprised not to find evidence of
ovarian inflammation. As a matter of fact, no lesion was
found sufficient to account for the extensive peritonitis.
It is quite possible that with greater experience I might
have been able to recognise and remove something of
greater pathological importance than a few subperitoneal
cysts and a tube distended with serum. For I know of
no operation in which experience is more helpful than in
this. For several years this patient had been my faithful
attendant at my consulting rooms, and the disastrous result
of the operation, which I was most unwilling to undertake,
distressed me exceedingly.

The three following cases, which also occurred before
I left Manchester, were fortunately more successful.

CASE 4. *Recurrent pelvic peritonitis commencing shortly
after marriage three years ago; constant pain for two years;
inability to work; small, fixed swelling on right side of*

uterus; abdominal section; chronic salpingitis of both sides; small suppurating ovarian cyst on right; left ovary adherent, otherwise normal; both tubes and both ovaries removed; recovery; complete disappearance of pain.—Mary B—, aged 25, married, was admitted into St. Mary's Hospital, Manchester, April 20th, 1886, complaining of severe pain on the right side of the pelvis and less severe pain on the left, also of a bearing-down sensation, worse after walking and at the menstrual periods. The symptoms commenced a few weeks after her marriage three years ago. Two years ago she was in the hospital under my care for some weeks, and left greatly improved. On resuming her household duties, however, she broke down again at once, and for two years the pain has now been constant, entirely unfitting her for work. She has never been pregnant.

The uterus is normal in size, mobility, and position. In the right posterior quarter of the pelvis is a mass about the size of a small orange, separated from the uterus by a distinct sulcus.

The general health is fairly good; the temperature normal. There has recently been some loss of flesh.

The diagnosis was dilated right tube. Abdominal section was performed April 30th.

The pelvic viscera were densely matted; a coil of intestine had become firmly adherent to the bladder. Both Fallopian tubes were thickened, each being half an inch in diameter. The right ovary was enlarged, its length being three inches. On section it was seen to contain two main cysts, one an inch in diameter, the other two inches. The larger cyst was full of pus. The left ovary was normal, but universally adherent. Both tubes and both ovaries were removed. A drainage-tube was inserted and retained for forty-eight hours. Menstruation commenced on the third day, and lasted until the seventh. Pain on movement of the right leg was complained of on the third day. Next day it was worse, but from that time it gradually diminished and eventually disappeared. On the sixth and seventh days there were hallucinations of sight on closing the eyes; these did not continue. The sutures were removed and an enema of olive oil was given on the sixth day; the bowels acted freely on the seventh. The temperature during convalescence never exceeded 100° F., and the patient was in due course discharged well. Six months afterwards she presented herself looking stout and well. The pain had entirely disappeared.

On October 26th, 1892, in reply to some inquiries, I received from the patient's medical attendant a letter, from which the following is an extract:—" The pain she had in the right iliac region has not troubled her since the operation. The pain in the left hip continued very constant until about two years ago, but since then she feels it only after a day's washing. She had rather a severe flooding about six months after the operation, and menstruated three or four times after that at irregular intervals. She has not menstruated now for two years. She has a continuous yellow discharge. She says she never was very strong, and at present considers herself as well as ever she was. The operation has certainly converted her from a chronic invalid into a woman capable of performing her household duties."

CASE 5. *Pain and tympanitic swelling in the lower part of the abdomen, commencing with an acute attack ten weeks before admission; after two months' rest and treatment pain diminished, but swelling increased; abdominal section; large abscess in peritoneal cavity, extending deeply into the right side of the pelvis, and shut off by adhesions; cavity emptied, washed out, and drained; purulent discharge for several months; rapid improvement of general health, and ultimately complete recovery.*—M. E. B—, single, aged 21, a weaver, was admitted into St. Mary's Hospital, Manchester, on April 12th, 1887, with swelling of the lower part of the abdomen, and complaining of pain, especially at the bottom of the back. The pain and swelling commenced ten weeks previously, at the end of a menstrual period. She had not menstruated since.

The lower half of the abdomen was uniformly distended; there was no fluctuation, and the percussion note was tympanitic throughout. No distinct tumour could be felt. The uterus was of normal size, its mobility impaired. Nothing could be made out as to the condition of the uterine appendages. After two months' rest in bed the size of the abdomen had rather increased than diminished. A distinct ridge could be felt running transversely across the abdomen a little below the umbilicus.

Abdominal section was performed June 8th, 1887. On opening the peritoneal cavity the omentum was found adherent to the anterior abdominal wall, and tacked down to the pelvis along its entire breadth. With much difficulty the right border of the omentum was separated and raised; it was then found that all the pelvic viscera were matted together by adhesions. In separating these the finger passed through a very friable membrane into a cavity, from which there escaped a quantity of thin sanious pus, mixed with flakes of lymph. The opening was enlarged, and the fluid soaked up, as it escaped, by means of sponges. The finger was then introduced within the abscess cavity, which dipped in the most irregular manner here and there amongst the viscera, and was evidently a portion of the peritoneal cavity shut off by adhesions. It extended a considerable distance upwards into the abdomen and downwards into the right side of the pelvis. The bladder formed part of its anterior wall. The cavity was washed out with warm water; the edges of the abscess sac were secured on each side, as well as their friable character permitted, to the edges of the lower part of the abdominal incision, and the upper part of this incision was closed. A drainage-tube was left in the sac. The uterus and appendages were not made out. There was some rise of temperature during the first week, the highest record being 101·8° F. at 2 a.m. on the 11th June (fourth day). On the third day the patient passed flatus through the rectal tube and was able to dispense with the catheter. Menstruation commenced the same day and continued until the sixth day. On the fourth day a discharge of offensive pus took place. The discharge soon lost its offensive character, but its quantity was for some time considerable. In the meantime the patient's health rapidly improved. In a fortnight she was sitting up, and on July 23rd she was allowed to go home for a few days. She was readmitted on August 17th, and as she became very useful as a ward help she was kept under observation for three months. There was still some purulent discharge from a small sinus when she left the hospital; this continued for some time, and finally ceased. I saw her in August, 1892, five years after the operation. She was then in excellent health, and was menstruating regularly. She had been married two years.

CASE 6. *Metrorrhagia and pain in the abdomen with bearing down, commencing two months after marriage; obscure retro-uterine swelling reaching to umbilicus, with increasing pain and tenderness and occasional rise of temperature; rest and hospital treatment for nine months without relief; abdominal section; large intra-peritoneal abscess; drainage; prolonged suppuration; recovery.—* Eva J—, aged 23, married, was admitted into St. Mary's Hospital, Manchester, on January 19th, 1887, complaining of irregular hæmorrhage and a sensation of bearing down. The symptoms dated from a few weeks after her marriage, which took place six months ago. She attributed them to having bathed in the open sea during menstruation. Three months ago some swelling of the lower part of the abdomen was observed, and she was thought to be pregnant. She had been kept in bed for some weeks previous to her admission.

On admission the abdominal walls were tense, but no definite tumour could be made out. There was dulness on percussion from pubes upwards to within an inch of the umbilicus. The uterus was normal in length, position, and mobility. She had an attack of pain in the hypogastrium on the 23rd of January, and was treated with poultices and the hot douche. She left the hospital relieved on March 5th, and was readmitted July 12th. Her general health had greatly improved, and the bearing-down sensation had almost disappeared. The menstrual flow had taken place regularly. She was examined under an anæsthetic on July 18th. Behind the uterus, which was normal, there was an obscure swelling rising into the abdomen nearly as high as the umbilicus. She went home again on the 23rd July, and was once more admitted on September 22nd, having become worse ever since leaving the hospital. She had suffered much more abdominal pain, the size of the abdomen had increased, and menstruation had been irregular, the intervals varying from three to five weeks. The temperature was raised, the appetite poor, and the patient was incapable of the least exertion.

The abdomen was swollen and tender, the muscles of the abdominal wall rigid. On bimanual examination a large fluctuating swelling could be felt behind the uterus, filling up the retro-uterine pouch and rising into the abdomen nearly to the umbilicus. The right lateral fornix was depressed by a firm swelling. No decided dulness on percussion, but the hypogastrium and part of each iliac region were duller than the rest of the abdomen; the flanks were resonant.

Abdominal section, October 12th.—Immediately beneath the abdominal wall, and adherent to it, was a swelling with a covering of what appeared to be peritoneum. During the separation of the adhesions the wall of the swelling was slightly torn, and some pus oozed out. The opening was enlarged, and about 20 fl. oz. of slightly fetid yellowish-green pus escaped, along with some lymph-flakes. The fingers were now passed into the abscess-cavity, which was found to be very extensive. It passed upwards above the level of the umbilicus, and dipped down into the pelvis. On the right side a prolongation extended to the pelvic floor. The uterus and appendages were not made out. The inner surface of the abscess wall was rough in places, but for the most part smooth and uniform. The edges of the opening were secured to the edges of the middle portion of the abdominal incision, and the incision, above and below, was brought together by silkworm gut sutures. A glass drainage-tube was inserted into the cavity and retained there for seventy-two hours, an india-rubber tube being then substituted.

Convalescence was very slow. The discharge was profuse, and as it became offensive the cavity was washed out daily with a solution of potassium permanganate. By the 5th of November the general health had begun to improve, and the amount of discharge from the wound to diminish. When she went home on the 10th of March, 1888, there was still a copious discharge from the sinus. which continued for some time. When I last heard of her, in July, 1892, four years and three quarters after the operation, she was perfectly well.

It is, to my mind, certain that in each of these three cases (4, 5, and 6) it would have been better to operate earlier. In none of them did the patient derive the least benefit from the delay. On the contrary, I believe that, had the abdomen been opened when the patients first came under observation, there would have been much less suppuration subsequently, and convalescence would have been far less prolonged. It is the experience derived from such cases as these, and from some others that will be related presently, that has convinced me of the general inexpediency of delay. If surgical relief is to be given, the more prompt that relief the better. In Case 4 two years were wasted, in Case 5 two months, and in Case 6 nine months, not to speak of the additional waste of time involved in the prolonged convalescence.

I now pass on to the cases that have occurred to me since I removed to London. The first of these, Case 7, is one that had been in the ward for some weeks under the care of my predecessor.

CASE 7. *Pain in left iliac region sixteen months; swelling twelve months; amenorrhœa six months; obscurely fluctuating tumour pushing uterus to right; severe illness with wasting and pyrexia; abdominal section; caseating abscess emptied and drained, edges secured to abdominal incision; rapid improvement in health, but sinus persistent, discharging muco-pus; sinus dissected out twenty-one months after operation; found to consist of left Fallopian tube; recovery; small sinus remaining.—*
E. F—, aged 25, single, a servant, was admitted into Adelaide Ward, St. Thomas's Hospital, under the care of Dr. Gervis, on February 13th, 1888, complaining of a swelling in the left iliac region, accompanied with constant pain and fever. The pain commenced in November, 1886, and the swelling was noticed in February, 1887, being then equal in size to a hen's egg. Menstruation, after gradually becoming scanty, ceased in July, 1887.

On admission she was very ill. Her temperature, usually ranging between 99° F. and 101° F., occasionally reached 102° F. and 103° F. She was losing flesh, and was in constant pain. There was a tense, hard, obscurely fluctuating tumour, causing a slight prominence in the left lower fourth of the abdomen. There was dulness on percussion over it. It was closely connected with the uterus; it reached in height from the pubic ramus to within half an inch of the umbilicus, and in width from the left lateral wall of the pelvis to an inch and a half beyond the middle line of the abdomen on the right.

When I came on duty at the end of March, the account given to me was that the patient had not improved during the six weeks she had been in the hospital; the swelling and pain had not diminished, and the loss of flesh and pyrexia had been continuous. I accordingly determined to make an exploratory incision.

Abdominal section was performed on the 5th of April, 1888. On opening the peritoneal sac some ascitic fluid and transparent jelly-like material escaped. The tumour was attached to the uterus (which was pushed over to the right), and was covered with peritoneum. There were no adhesions in front or behind. A trocar was inserted and 3 fl. oz. of pus withdrawn. The opening was then enlarged to the length of an inch and a half, and the finger inserted. The wall of the abscess cavity was $\frac{1}{4}$ in. thick, and lined, on its roughened inner surface, with caseous material, of which as much as possible was pressed and scooped out. After washing out the cavity with hot boracic solution, and the peritoneum with simple hot water, the wall of the abscess was stitched to the edges of the abdominal incision, the rest of which was closed by sutures of silkworm gut. An india-rubber drainage-tube was inserted into the cavity.

Next day the temperature rose to 102°, and the pulse to 150. On the third day the temperature ranged from 98·6° to 101·2°; on the fourth, from 99° to 100·4°; on the fifth, from 98·6° to 101·6°; and on the sixth, from 98·4° to 99°. After that it was uniformly normal.

There was a copious discharge of pus, and three weeks after the operation a quantity of cheesy material was cast off with the discharge. After the first five days the patient's general condition quickly and permanently improved. She gained flesh, and was able to sit up in bed at the end of a fortnight. In a month the tumour had contracted, its upper limit being 2 in. below the level of the umbilicus.

She left the hospital, on the 12th of July, stout and well, but still wearing the drainage-tube. The sinus was 2½ in. long, and about 3 fl. oz. of muco-pus escaped during each twenty-four hours. She had menstruated once.

On September 18th, 1888, she presented herself at the hospital. Her condition had still further improved. She still wore the tube; the discharge was now slight. She had menstruated twice since leaving the hospital.

At the beginning of 1890 the patient was still wearing a drainage-tube, all attempts to discard it, even with curetting of the sinus, having failed. This fact, together with the continued presence of mucus in the discharge, convinced me that the abscess was not in the connective tissue of the broad ligament, as was thought at the time of operation, but in a cavity lined by mucous membrane. By stitching the edges of the abscess wall to the edges of the abdominal incision, a fistulous communication had evidently been established between this cavity lined by mucous membrane and the exterior. It seemed to me highly probable that the case was one of pyosalpinx, and that I had unintentionally performed the operation of salpingostomy.

I therefore readmitted the patient, and on the 14th January, 1890 (a year and nine months after the operation), the sinus was carefully dissected out. It was found to consist of the left Fallopian tube, thickened, but no longer dilated, running directly forwards from the left cornu of the uterus, which had become twisted half round on its vertical axis, so that its anterior surface looked to the right, and its posterior to the left. The tube was removed close to the uterus, the exposed mucous membrane in the stump being cauterised by a heated iron skewer. The normal right tube and ovary were felt behind the uterus.

The last time I saw this patient, viz. on July 25th, 1891, she was strong and well, although there was still a very slight muco-purulent discharge from the old sinus.

In a letter I received from her December 14th, 1892, she told me she was about to be married. Menstruation was regular, generally painful and somewhat profuse. There was still a slight discharge from the sinus.

The lesson to be learned from this most interesting case is not to be satisfied with half-measures. Regarding the case as one of abscess in the broad ligament, I did not attempt to do more than empty and drain it. The sequel showed that the whole cyst should, if possible, have been removed. As to the nature of the abscess, the presence of a quantity of caseous material points strongly to tubercle. No microscopic examination, however, having been made, the tubercular character of the mischief is necessarily conjectural. There is strong reason for believing that all cases of pyosalpinx in the virgin (and this patient had the physical signs of virginity) are tubercular in their character.

CASE 8. *Illness of twelve months' duration; tense fluctuating swelling above pubes; pain in left iliac region; pyrexia and wasting; abdominal section; pelvic peritonitis, with encysted collection of serum; fluid removed; immediate relief of symptoms; recovery.*—Alice L—, aged 20, a widow, was admitted to St. Thomas's Hospital May 12th, 1888, with symptoms of pelvic peritonitis, and a supra-pubic swelling which had not hitherto been noticed.

She had given birth, a year previously, to a stillborn child at about the seventh month of pregnancy, and had suffered from pain in the left iliac region ever since. She had been unable to work, but had not been confined to bed until quite recently. She was now thin, pale, and ill; her temperature was 102·6°, her pulse 114. Above the pubes was a distinctly fluctuating swelling, three inches in its vertical measurement, and extending three inches to the right of the middle line, and a little less to the left. It was tender to the touch, dull on percussion, and immoveable. The uterus was fixed, displaced somewhat to the right, and of normal length. Above the vaginal roof on the left side, a tense brawny swelling could be felt. The fundus of the bladder was situated an inch above the pubes. The swelling was thought to be an abscess.

Abdominal section, May 21st.—The contents of the pelvis were completely roofed over by adherent omentum. On separating the omentum the swelling was exposed to view. A bladder sound was introduced, and showed the fundus of the bladder to reach only to the lower angle of the abdominal incision. A small trocar was passed into the swelling, and a little straw-coloured serum escaped. The opening was enlarged by means of the finger and the cavity explored. It was found to be lined by peritoneum and to be very irregular, dipping here and there amongst the pelvic viscera. It was bounded by the uterus on the right, and by the left broad ligament in front and to the left. A glass drainage-tube was inserted and the abdominal wound closed.

The temperature, which during the week preceding the operation had ranged from 99° to 100·4°, fell at once to normal, and only once reached 99° during convalescence. A little suppuration took place from the tube-track at the beginning of June, but only lasted a few days. On the 6th June the patient was able to sit up. On the 12th the uterus was found still slightly displaced to the right, and a small fluctuating swelling was detected above the vaginal roof on the left side. On the 19th this swelling had disappeared, and the uterus was nearly in the middle line. The patient was sent to a convalescent home on the 20th, and on the 18th of July she returned, looking and feeling perfectly well. She had gained flesh, had a healthy colour in her cheeks, and was in the highest spirits.

In September, 1892, she was readmitted. Having remained well and at work for four years and a quarter, she had a sudden attack of pelvic pain a week before admission. A hard irregular mass was found in the right posterior quarter of the pelvis. Abdominal section was again performed, and the uterine appendages on the right side were removed for chronic inflammatory disease.

This case was a good illustration of the effects of tension. Encysted collections of serum in the pelvis produce no symptoms unless there is tension, when they give rise to severe constitutional disturbance, and may easily be mistaken for pelvic abscess. Indeed, I do not know how the two conditions can be distinguished. The diagnosis is of the less importance, however, as the indications for treatment are the same in both. The reason for the swelling making its appearance above the pubes was that Douglas's pouch was nearly obliterated by adhesions.

An outline of the next case was published in the 'British Medical Journal' for July 20th, 1889. The parts removed at the operation and at the autopsy had already been exhibited at a meeting of this Society, along with a coloured drawing which the Council did me the honour to publish.

CASE 9. *Gonorrhœa; right hydrosalpinx; abdominal section; removal of distended tube and adjacent ovary; death from acute peritonitis in fifty-six hours; autopsy; pus in the pelvis, in the left tube, and in remains of right tube; perforating ulcer of intra-uterine portion of both tubes, cicatrising on left, more recent on right.*—Mary C—, aged 19, single, until recently a prostitute, was admitted into Magdalen Ward in May, 1888, suffering from gonorrhœa, and transferred to Adelaide Ward, August 20th, 1888, on account of pain in the left iliac region, supposed to be due to ovaritis.

At the latter part of 1887 she had a yellow vaginal discharge, with pain in both iliac regions, lasting for eleven weeks. After being better for a month these symptoms recurred in March, 1888, when a swelling developed in the left side, which varied in size from time to time. On being admitted to Magdalen she complained of pain only on the left side; she had a thick purulent vaginal discharge, which was most profuse when the swelling was less marked, and less so when it became hard and well defined. Sometimes the discharge was blood-stained. There was no pain on micturition. During her stay in Magdalen she had an attack of very severe pain in the left side, with a high temperature and extreme prostration, thought at the time to be due to acute ovaritis.

On admission to Adelaide Ward there was discovered a slight lateral displacement of the uterus to the left.

Lying behind and to the right of the uterus was a not very tense, smooth, oblong swelling, equal in size to an egg, and giving a sense of fluctuation. This was diagnosed as a hydrosalpinx of the right tube, the tube having become occluded at its fimbriated extremity and bent upon itself, so that the outer distended portion lay behind the inner portion and the uterus. There was still a purulent discharge from the vagina. On the evening of September 12th, after having been examined bimanually, the patient was sick and complained of acute pain in the right iliac region. The temperature rose to 103·4°, and the pulse to 134. The patient looked ill and somewhat collapsed. The right iliac region was swollen and tender. It was thought that the swollen tube must have been a pyosalpinx that had ruptured, and it was decided, if the symptoms did not improve, that the abdomen should be opened. Next day, however, the patient was much better, and the temperature fell to what it was before the attack. The swelling and tenderness gradually disappeared. On September 22nd I ventured, for the first time since the attack, to make a vaginal examination. The result was that I found the retro-uterine swelling unaltered, or, if anything, a little fuller and more tense.

On October 18th abdominal section was performed for the removal of the dilated tube, which the illness of the previous month led me to regard as a source of danger. The dilated tube was pyriform in shape, measuring three and three quarter inches in length, two inches and a quarter in breadth at its widest, and an inch and a quarter at its narrowest part. The broadest part was at the fimbriated extremity, which was closed. The dilated portion was confined to the outer part of the tube, and was lying behind the uterus, the undilated part of the tube being bent upon itself. There were no adhesions about the swollen tube, and it was removed, along with the adjacent ovary, without difficulty. The contents of the dilated tube were serous. The left tube felt as though it contained hard nodules in the substance of its walls; the left ovary was adherent. The left appendages were not removed.

The patient died of septic peritonitis fifty-six hours after the operation.

At the necropsy (made by Dr. W. B. Hadden) there were found some recent peritoneal adhesions in the lower part of the abdomen; a small quantity of thick pus was found in the pelvis. There were two black spots on the peritoneal aspect of the fundus uteri, one at each cornu.* The tissues beneath were disorganised. A band-like process of great omentum passed to the gangrenous spot on the left side, and was firmly adherent there. The cavity of the uterus was of average size; the mucous membrane was coated with fluid blood (menstrual?). On opening the remains of the right Fallopian tube from within, the first half of the intra-uterine portion was normal, the second or outer half was ulcerated, and a perforation, seven millimetres in length, existed on its upper surface corresponding to the gangrenous spot already described as existing on the right cornu of the uterus. From the outer border of the uterus to the point where the tube had been divided the lining membrane appeared healthy. There was a little pus lying in the tube. The left tube was a little dilated, especially at its distal part, which contained some pus. On opening the intra-uterine portion of the tube, the inner half of that portion was healthy in appearance; the outer half was either occluded, or at any rate so constricted that the finest wire could not be made to pass. Between the constriction and the black spot on the peritoneal surface the tissues were softened and of a deep red colour. No communication could now be detected between the interior of the tube and the peritoneal cavity. Beyond this were two hard nodules (gummata?) which, on section, were seen to be pale circumscribed masses of exudation, completely surrounding the mucous membrane. The left ovary was of normal size and much softened.

* See coloured plate in the 'Trans. Obstet. Soc.,' vol. xxx, p. 406.

This case, so far as I know, is unique. It shows to what unsuspected risks patients suffering from gonorrhœal salpingitis are exposed. If ulceration can take place in the intra-uterine portion of the tube to such an extent as to destroy the whole thickness of the uterine wall, and, perforating the peritoneal coat, allow the purulent contents of the tube to discharge themselves into the peritoneal cavity, it is obvious that even removal of the tubes would not suffice to avert the risk. Fortunately this portion of the tube appears to be ulcerated so rarely that, for practical purposes, we may leave this danger out of account. Besides, the case before us shows that perforation is not necessarily fatal. There can be little doubt that the alarming symptoms that supervened whilst the patient was in Magdalen Ward, when it will be remembered all the suffering was on the left side, mark the time when the perforation of the left tube occurred ; and that the equally alarming symptoms that occurred after an examination in the month of September marked the precise moment when the perforation took place on the right side. On both these occasions the patient became collapsed, and was for some hours in extreme danger, but the peritoneum of this young and robust subject proved equal to the emergency, the extravasated matters became absorbed, and a friendly band of omentum sealed up the aperture. The hydrosalpinx, which was the only lesion discovered or discoverable on vaginal examination, was, of course, a mere retention-cyst produced by the closing, during one of the attacks of pelvic peritonitis, of the fimbriated end of the tube. In itself the lesion did not justify an operation, but it was evident from the recurrent attacks of acute pelvic inflammation that there was something more than hydrosalpinx. Hence I decided to open the abdomen. I did not, however, even during the operation discover anything beyond the hydrosalpinx. The black spots at the uterine cornua were concealed from view by bands of omentum, and the left tube, in external appearance, was as nearly as possible normal. With regard to the fatal result of the operation, I am quite unable to offer an explanation. I instituted a most minute inquiry as to the possibility of any antiseptic precaution having been overlooked, but without result.

Two other points I wish to call attention to before I pass on, namely, (1) the fact that in the same tube a collection of serum may exist at one end, and a collection of pus at the other; and (2) the fact that rupture of the Fallopian tube may take place at a part where there is no appreciable dilatation. To this latter point Dr. Lewers has already directed attention (see 'Trans. Obst. Soc.,' vol. xxvii, p. 298).

CASE 10. *Recurrent pelvic peritonitis; constant pain more or less severe, and general feeling of illness for last fifteen months; fluctuating tumour above pubes; abdominal section; removal of pedunculated retro-peritoneal cyst with two daughter-cysts, the latter suppurating; death on eighth day; autopsy: small quantity of pus in pelvis; partial obstruction of small intestine at site of old adhesion.*—Sarah T—, aged 32, single, a dressmaker, was admitted into Adelaide Ward December 13th, 1888. Five years ago, when over-worked as a teacher, she caught cold (not during a menstrual period), and had a severe illness with much abdominal pain, incapacitating her for six months. After she came to reside in London she felt well until the autumn of 1887, when she had a similar attack; a third took place three months before admission. Since that time the abdominal pain has been constant, sometimes severe, sometimes slight. Menstruation has been regular and painless throughout.

On admission, patient looked thin, sallow, ill, and tired. She was of a highly nervous temperament and unusually intelligent. She complained of some fulness at the lower part of the abdomen, but was not aware of the existence of any tumour.

The abdomen was rendered very slightly prominent by

a rounded fluctuating tumour, situated almost centrally and reaching from pubes to umbilicus, a distance of 6½ inches. It extended 3 inches to the right and 2½ inches to the left of the middle line; it was dull on percussion. The uterus was normal in size and consistence, and was pushed to the left side, the sound passing with difficulty after being slightly bent. The urine was loaded with lithates. Temperature ranged from 98·6° to 101°.

Abdominal section December 20th, 1888. The omentum was adherent to the cyst, and there were some recent adhesions to the anterior abdominal wall, especially on the right. After these had been separated, the cyst, which was covered by peritoneum, was tapped. Thirty fluid ounces of dark brown fluid (proving on microscopical examination to be altered blood) were removed, with some thick, grumous, flaky material, and, towards the end, some pus. The cyst-wall was very pliable, and gave way in all directions on the slightest manipulation. The remaining adhesions were then separated; they were very numerous, firm and vascular, and involved intestine, mesentery, and parietal peritoneum. The pedicle, which could not be brought into view, was secured with a single ligature and divided. The cyst consisted of one main and two daughter cysts; the latter had both been in a state of suppuration, and had burst into the main cyst during the operation. The right Fallopian tube was not seen. The uterus and the left ovary and tube were matted densely together by old adhesions; Douglas's pouch was obliterated by adhesions. The peritoneum was flushed, a glass drainage-tube was inserted into the right side of the pelvis, and the wound was sutured.

At 9.30 a.m. the following day there had been no sickness; the tube was removed.

On the third day (December 22nd) patient became very restless, and the pulse rapid, flickering, and uncountable. There was no pain.

On December 23rd the condition was very alarming : extremities cold, bowels acting involuntarily, respiration embarrassed, slight distension of abdomen ; no pain and no sickness. Towards evening patient appeared to be moribund. At 4 a.m. on the 24th she was apparently dying, when suddenly she sat up and asked to have the pillow changed. During that day she remained a trifle better, but continued very nervous and irritable. The bowels were relaxed, the motions passing unconsciously. She continued in much the same state and quite conscious up to 4 a.m. on the 27th, when she lost consciousness, and she died at 8 a.m.

The highest temperature on the day after the operation (viz. on December 21st) was $100·6°$; on the 22nd, $99·8°$; on the 23rd, $100·4°$; on the 24th, $99°$; on the 25th, $97·6°$; after which it rose once to $99·2°$, but was generally sub-normal.

Autopsy (by Dr. H. P. Hawkins).—Omentum firmly adherent to wound; a small collection of pus under its lower end. Lower end of omentum, passing through coils of small intestine, was firmly attached by an old adhesion to the back of the pelvis, by the side of the rectum and transverse colon, which latter, collapsed and empty, had been drawn out of position by the omentum. Superficial coils of small intestine much distended with gas ; some injection of vessels along lines of contact, but only a few shreds of lymph. There was a little blood-stained fluid free in the lateral parts of the peritoneal cavity. The coils of intestine that lay in the pelvis were acutely inflamed, and adherent to each other by soft, recent blood-stained lymph. Between the coils on the left side was a collection of about half a fluid drachm of green viscid pus. On removing the intestines the floor of the pelvis seemed levelled by adhesions and deposit of inflammatory material, there being no sign of bladder, uterus, ovaries, or broad ligaments. On this floor lay two or three fluid ounces of viscid greenish pus, without odour. The uterus and adnexa were scooped out. The left ovary and tube were adherent on all sides, and lay behind the uterus and left broad ligament. The right ovary and tube were also found amidst a mass of adhesions. The remains of the pedicle, with ligature attached, were found projecting from the peritoneum, covering the lower part of the back of the corpus uteri. The uterus itself was normal. The tumour removed was evidently a cyst underlying the peritoneum. There had been no secondary hæmorrhage. Where a coil of small intestine crossed the right side of the pelvic brim, it was firmly attached to the psoas by old adhesions, causing partial obstruction. Meckel's diverticulum and the appendix vermiformis were normal. Left pleura completely and firmly adherent, the lung being torn during removal. Right pleura adherent over apex. No fluid in pleuræ. A few caseous or partially calcified nodules at apex of left lung. Some hypostatic basal congestion. Anterior surface and edge of right lung extremely emphysematous; caseous nodules at apex, rest healthy. Heart and other organs normal.

Of the two possible causes of death in this case, viz. the partial obstruction of the small intestine and the septic peritonitis, the latter seems the more probable. There is little doubt that the source of infection was the purulent matter that escaped from the cyst during the operation, a portion of which must have remained in spite of the flushing. Any way, I determined not again to rely upon flushing alone in the event of a similar accident, but to sponge carefully whether I flushed or not.

With regard to the precise nature of the cyst I do not feel able to offer an opinion. It was not connected with either of the tubes, the ovaries, or the broad ligaments. It was covered by peritoneum, and was attached by a distinct pedicle to the back of the uterus, an unusual position for a cyst of this character.

The case, though an exceptional one, is included in this series because the patient sought relief, and the operation was undertaken, on account of the recurrent attacks of pelvic peritonitis.

CASE 11. *Chronic salpingitis and chronic pelvic peritonitis; hæmorrhage from both Fallopian tubes, forming intra-peritoneal hæmatocele on each side of the pelvis, encysted amongst old pelvic adhesions and embraced by the expanded fimbriæ of the tubes; abdominal section; removal of blood-clots and both tubes; death on ninth day from acute nephritis.*—The patient, a married woman aged 32, had recovered well after each of her four confinements, the last of which took place two years and seven months ago. Eighteen months ago she had a miscarriage, followed by an illness of eight weeks' duration. There had been two early miscarriages since, the last one twelve weeks before admission. The patient dated her illness from that time. She had suffered during the past month from pain in the back and in the right iliac region, and latterly there had been pain during micturition and defecation.

Nothing abnormal could be detected in the abdomen. Behind and to the right of the uterus, which was of normal size, fairly moveable, and situated slightly to the left, was a smooth, firm, elastic, immoveable swelling, which occupied the right posterior quarter of the pelvis, and extended an inch to the left of the middle line. The left fornix was narrowed. High above it could be felt an obscure swelling, tender on pressure. I have unfortunately no note of the diagnosis. All I can say on this point is that I was not prepared to find that the main swelling was a blood-clot.

On opening the abdomen, a rounded solid tumour, apparently continuous with the right Fallopian tube, was found occupying the retro-uterine pouch, and extending outwards to the right pelvic wall. From the outer side of the swelling the tube curved forwards and inwards to the right cornu of the uterus. The mass was fixed by extremely firm adhesions to the pelvic walls and to the rectum. On the left side a similar but much smaller mass was situated behind the left broad ligament. The body of the uterus was free and fairly moveable. With the exception of the rectum, the intestines were not involved. It was evident there had been old pelvic peritonitis, and that amongst the matted tissues were two solid tumours, one on each side, that on the right being the larger. The masses were with extreme difficulty separated by the fingers. The larger tumour was first brought into view. It consisted of a firm blood-clot, equal in size to a hen's egg, and of a more or less globular shape, and was embraced by the expanded fimbriæ of the right tube. The tube itself was thickened, empty, and undilated, and was bent backwards upon itself. The broad ligament was also much thickened. The ovary was not seen. The tube was removed with the tumour. The smaller mass, on the left side, also consisted of firm blood-clot, laminated and partly decolourised. Like its fellow, it was embraced by the fimbriæ of the corresponding tube. The tube and blood-clot were removed. The ovary, white and shrivelled, was firmly adherent to the pelvic wall, and was not removed.

The patient was much collapsed after the operation. Next day the urine was found to contain a trace of albumen. The quantity of albumen increased, and the urine became scanty and smoky. Death took place on the ninth day, the temperature, except on the day following the operation, having been uniformly under 100°.

At the autopsy the kidneys were intensely hyperæmic, and generally showed evidence of acute nephritis. The retro-uterine pouch was occupied by two feet of small intestine, which had contracted slight adhesions. On removing them the pouch was seen to be lined with a thin layer of firm stratified blood-coagulum, one sixth to one eighth of an inch in thickness. No fluid blood was present; no pus; no general peritonitis; no serous effusion; no obstruction or strangulation of bowel; no visceral injury. The ureters also were normal and uninjured. The post-mortem examination was made by the late Dr. Gulliver. He concludes his report by stating that, in his opinion, the cause of death was acute nephritis, the parts concerned in the operation appearing to be as healthy as could be desired.

I believe a complete diagnosis before operation was in

this instance impossible. The hæmorrhage appeared to have been secondary to inflammatory changes in the tubes, and the clots assumed their misleading shape and position from being imprisoned amongst old pelvic adhesions. The cause of death was, so far as my experience is concerned, an unusual one after these operations.

CASE 12.—*Recurrent pelvic peritonitis extending over five years; abdominal section; chronic inflammation of both Fallopian tubes; small suppurating cyst of left ovary; removal of both ovaries and both tubes; uninterrupted recovery.*—The patient (S. A. W—), an unmarried girl aged 22, had been delivered of a full-term child at the age of fourteen. Two years afterwards she began to suffer from pain and swelling in the lower part of the abdomen, and a yellow vaginal discharge, for which she underwent a course of treatment in the Bridgnorth Infirmary. Two years later she had a recurrence of the symptoms, and again became an inmate of that institution. Five months before admission she is said to have caught cold during menstruation; an attack of shivering occurred, and the flow ceased for a few days. In two months from that date she sought admission into the Bridgnorth Infirmary for the third time; she remained there, in bed, for six weeks, and was then transferred to St. Thomas's Hospital.

She was pale but not emaciated. She complained of pain in the back and in the right iliac region. The uterus was normal in length, fixed, and strongly flexed to the right. Extending from the uterus to the left pelvic wall was a thick, smooth, hard, elastic, slightly moveable mass, the outer extremity of which was on a level with the anterior superior spine of the left ilium, and three quarters of an inch internal to it. She had come up to London with the view of undergoing an operation, but as the pyrexia, which had been a very marked symptom up to the time of leaving Bridgnorth, disappeared from the moment of her arrival at St. Thomas's Hospital, I thought the swelling might be merely a hydrosalpinx surrounded by firm adhesions, and determined to watch the case a little before proceeding to operate. She was accordingly kept in bed for six weeks. At the end of that time, the swelling being no less, and the patient, though less anæmic, being still unable to move about, it was decided to make an exploratory incision. Only on two occasions (February 11th and March 1st) had the temperature exceeded the normal during the whole six weeks.

On March 21st the abdomen was opened. On the left side was found a small ovarian cyst, $3\frac{1}{2}$ inches long by $2\frac{1}{4}$ inches wide, filled with thin flocculent pus. The left tube, thickened and dilated, was adherent to its surface. Cyst and tube were separated from their adhesions and removed. On the right side the Fallopian tube was found dilated, its walls œdematous, and its fimbriated extremity adherent to the floor of Douglas's pouch. The ovary was double its normal size and almost universally adherent. The tube and ovary were separated from their adhesions and removed.

Convalescence was uninterrupted, the temperature never exceeding 100°.

I had a letter about her in January, 1891. She was quite well, free from pain, and following her employment as a domestic servant.

This case tells its own story. I need not, therefore, detain you by comments upon it.

CASE 13. *Recurrent pelvic peritonitis and cellulitis; hard mass behind and to left of uterus, thought to be subperitoneal fibroids; great improvement under rest; readmission a year afterwards; exploratory incision; diagnosis confirmed; discharge of pus per rectum; abdomen reopened; deep-seated abscess opened, emptied, and drained; recovery.*—The patient, a married woman aged 39, had borne seven children and had had two miscarriages. After her last confinement, which took place twelve years ago, she was ill and feverish for two weeks.

On admission, May 23rd, 1888, she had been losing flesh and in poor health for twelve months, for the last four of which she had been suffering from abdominal pain and tenderness, worse after walking. A fortnight before admission she had had a sudden attack of acute pain, and the bowels had not acted for six days. The pain continued more or less up to her admission, and was accompanied with vomiting. She was a tall, strongly built woman, but pale, emaciated, and very ill. Above the vaginal roof, posteriorly and to the left, was a hard, tender, irregular mass. The cervix was fixed, and partially surrounded by induration.

The patient was kept in bed for a month and poulticed, her temperature being normal throughout. At the end of that time she had improved immensely, having regained flesh and lost her look of illness. The resistance and tenderness in Douglas's pouch had diminished, and the hard lump on the left side was more clearly defined. The case was thought to be one of subperitoneal fibroids of the uterus, with pelvic peritonitis and cellulitis. She remained in the hospital another month, still continuing to improve in her general condition, and was discharged on the 2nd August, 1888.

She remained well until October, 1888, when she again began to fail. Shortly after that she attended as an out-patient occasionally, and on July 17th, 1889, she was readmitted into the ward. Her general condition was much the same as when she was first admitted, fourteen months previously. There was no increase of temperature. A large, irregular swelling could be felt on bimanual examination extending from the uterus posteriorly, and to the left lateral wall of the pelvis. I was still inclined to the belief that the main swelling was a mass of subperitoneal fibroids, but there being an element of uncertainty about it, I suggested an exploratory incision, to which she eagerly consented

On August 2nd, 1889, I accordingly opened the abdomen. Behind and to the left of the uterus, was a smooth hard mass, quite immoveable, and covered by the peritoneum, to which coils of intestine were adherent. No fluctuation could be detected in it. The mass appeared to spring from or to be very closely attached to the left side of the uterus. The impression conveyed was that of a fibroid burrowing beneath the peritoneum. Under these circumstances the abdomen was closed without any attempt at further interference.

After the operation the bowels acted five times, and one of the motions was observed to contain a quantity of pus. It then transpired for the first time that yellow matter had from time to time been evacuated with the stools since the month of May. This threw a new light upon the case. It was now fairly certain that the mass, which had been thought to be a fibroid, was a thick-walled pelvic abscess, which communicated with the rectum, the size of the aperture being insufficient for complete evacuation. The temperature still remained normal. I proposed, however, in the light of the fresh facts which had come to my knowledge, to reopen the abdomen. I did so, four weeks after the former operation. The internal appearances were the same as on the last occasion. I now proceeded to pass a medium-sized trocar into the swelling (after having cleared it of adherent intestine, &c.), and withdrew an ounce or two of very offensive pus. I then removed the trocar and cannula, and enlarged the opening, by means of a scalpel, to a size sufficient to admit my finger, which passed deeply down into a smooth-walled cavity. The edges of the opening were then secured to the edges of the incision in the abdominal wall by two silk sutures on each side, and a 4-inch glass drainage-tube inserted. A second drainage-tube was passed down to the pelvic floor on the right side of the uterus, to drain the peritoneal cavity. The upper part of the abdominal wound was then closed by silkworm gut sutures in the ordinary way.

The patient made an excellent recovery. She had no vomiting and no rise of temperature from beginning to end. The drainage-tube in the peritoneal cavity was removed the morning after the operation. Within forty-eight hours of the operation all the sutures connecting the abscess-cavity with the abdominal incision were removed, and the glass drainage-tube used to drain the abscess-cavity was replaced by one of india-rubber. This was finally removed, on the fourteenth day, and on the twenty-eighth day all discharge had ceased.

The patient left the hospital on the 28th September stout and well, and has remained well ever since, except that she has a hernial protrusion at the lower part of the abdominal wound. There has never been seen the slightest stain of matter from the rectum since the day of operation.

I have described this case in some detail because our most useful lessons are learnt from our mistakes. I ought to have known there was pus in that pelvis from the recurrent peritonitis, which I now know to be a far truer test than the temperature. Even when the operation was concluded I felt unable to give an opinion as to the precise character and situation of the abscess. I have no doubt now, after larger experience, that it was either a pyosalpinx or a small suppurating cyst of the ovary, adherent to and covered in by an enormously thickened broad ligament. I have also no doubt that if I had to operate on the case to-day I should not be satisfied with emptying and draining, but should remove the diseased part, after separating it from its adhesions to the broad ligament and other surrounding structures. Had this been done, the hernia would in all probability have been avoided.

Not less interesting or less successful is the case that comes next in order.

CASE 14. *Recurrent attacks of pelvic peritonitis following gonorrhœa; great emaciation and inability to earn a livelihood; abdominal section; purulent salpingitis with intra-peritoneal abscesses; left tube removed; abscesses emptied and drained; acute pneumonia; recovery.*—A brief account of this case appeared in a paper published in the 'British Medical Journal' for December 27th, 1890, on the "Differential Diagnosis of Pelvic Inflammations," from which I take the liberty of quoting a paragraph or two. The patient was "a young woman, aged 28, with a worn, pale face, and wretchedly thin. She was admitted September 10th, 1889, complaining of severe pain in the lower part of the abdomen, and with a temperature of 101½°. She had been married five years, but had been separated from her husband for three years on account of his intemperance and cruelty, and during this time had had to maintain herself and her two children by dressmaking. Only on one occasion since their separation had she and her husband cohabited. This act of intercourse took place twelve months before her admission. Very soon afterwards she began to have a profuse yellow vaginal discharge. . . . In a few weeks she became too ill to continue at her work, and had to give up her home and go into the parish infirmary with her children. She came out in three or four months, but soon had to return. She again took her discharge and resumed her occupation. Her health, however, soon gave way again. She suffered great pain in the lower part of the abdomen, in the groins, and in the back, and eventually sitting became so difficult and painful that she had to relinquish her employment, and for some weeks before admission she had subsisted on the generosity of friends."

The uterus was pushed over to the left side and to the front by a tender, irregular mass filling the right side of the pelvis and Douglas's pouch. The diagnosis was gonorrhœal salpingitis with suppuration, and pelvic peritonitis.

"Abdominal section was suggested and readily agreed to. The operation was performed on September 14th, 1889. There was no general peritonitis, but the pelvis was occupied by a mass of adherent viscera, difficult to recognise and separate. The uterus, of normal size, lay in front and to the left." The left tube, much thickened, first ascended and then curved abruptly downwards and backwards, so that it lay mainly behind the uterus, where it was firmly adherent. At the angle of flexion it presented a distinct knuckle of enlargement. Its upper surface was free. Its lower surface was coated with old blood-clot, and formed part of the wall of a small abscess-cavity, from which, when opened, there welled up blood-stained serum, lymph flocculi, and pus. The cavity was intra-peritoneal. The broad ligament was much thickened, and both it and the swollen tube were so friable that when the tube had been separated from its adhesions, and was about to be removed, the ligatures placed around the broad ligament tore through. Some bleeding took place from the torn surface, and was arrested by four fine silk ligatures passed through the broad ligament, and tied over the cut surface. On being removed the tube was seen to be pervious throughout. A thin, purulent fluid exuded from it on pressure. Its walls were greatly thickened, as also was the mesosalpinx. Drawings of the tube were published in the paper already alluded to.

The right tube was less thickened but much distorted, and very intimately adherent to surrounding parts. Its direction was first forwards, then backwards and downwards, terminating behind the uterus. The cæcum and its appendix being apparently involved in the adhesions, the tube was not removed. It was separated, however, sufficiently from its adhesions to open up a second small abscess-cavity, distinct from that on the left side, and separated from it by a vertical septum. The contents of the two cavities were similar. A glass drainage-tube was inserted into each cavity after it had been well douched with hot boracic solution. The ovaries were not distinguished. The operation lasted an hour and a half.

"The patient had a severe attack of pneumonia after the operation, and there was a good deal of suppuration through the drainage-tube before the abdominal wound entirely closed, but she eventually made an excellent recovery, gaining flesh, and looking quite bright and cheerful. Before going out she complained of a vaginal discharge, which, on examination, proved to be due to a purulent inflammation of the urethra and nymphæ, and to a purulent cervical catarrh, for which she underwent the usual treatment before she left the hospital. The gonorrhœal origin of the pelvic inflammation was thus abundantly confirmed. The patient was able to be sent to a convalescent home on October 29th," after being seven weeks in the hospital.

I have recently been at some pains to trace her whereabouts, but without success, so that I am unfortunately not able to report her present condition, or to say whether the remarkable improvement effected by the operation has been maintained.

The case, as I have already remarked elsewhere, was a typical example of the class of cases that until recently were regarded as pelvic cellulitis, and treated accordingly.

CASE 15. *Recurrent pelvic peritonitis following gonorrhœa; fixed, tense, oblong swelling in right side of pelvis, with purulent endometritis; abdominal section; pyosalpinx on right side; prolapsed and adherent but otherwise normal ovary; right tube and ovary removed; recovery without rise of temperature; readmission for curettage of uterus; cure.—* The patient was a prostitute, aged 22. Two years ago she had a yellow vaginal discharge and a sore, followed by enlarged glands in the groin, and, later, by sore throat and blotches on the face. Four months ago she was seized with sudden and severe pain in the lower part of the abdomen, chiefly in the right iliac region, shooting down the right thigh and causing her to draw up the knee. She was feverish and kept her bed for two days. She vomited several times and had diarrhœa. There was a somewhat copious vaginal hæmorrhage, and irregular hæmorrhages have occurred from that time, especially after exertion and always after intercourse. A similar attack of pain with fever took place a fortnight after the first attack, and a third one two weeks ago. On each occasion she was in bed for about three days.

Patient is a healthy blonde, in good muscular condition. She has had more or less vaginal discharge, sometimes white, sometimes yellow, ever since the acute attack of gonorrhœa two years ago.

There is no abdominal swelling, but a feeling of resistance in the right iliac region. The vulva is normal, save for a stain, such as would be produced by silver nitrate, on the fourchette. Uterus is of normal size, displaced to the left and fixed. In the right posterior quarter of the pelvis there is a fixed, ill-defined, tense, oblong mass, which can be felt to bulge into the rectum anteriorly and to the right side. Nothing abnormal can be detected to the left of the uterus. At the bottom of Douglas's pouch can be felt a small cystic swelling, like an ovary. Temperature normal.

The diagnosis was right pyosalpinx, with prolapsed and adherent ovary.

The patient being willing to have an operation, the abdomen was opened on October 17th, 1889. The right tube and ovary were displaced behind the uterus and firmly matted to surrounding parts. The tube was enlarged, tortuous, and distended; its closed fimbriated end, measuring an inch in diameter, was adherent to the bottom of Douglas's pouch. The tube was studded with a number of subperitoneal cysts; one at the outer end had been felt on vaginal examination, and had been mistaken for a small prolapsed cystic ovary. At the angle of flexion, near the uterine end of the tube, the adhesions were very firm to the vermiform appendix and other parts. The ovary was slightly enlarged, and contained a number of cystic dilatations, some of them being filled with serum, others with altered blood.

The right tube and ovary were removed. The tube was found to be distended with pus. The left appendages appeared to be normal.

The patient made an uninterrupted recovery, her highest temperature being 99·4°. She left the hospital on the seventeenth day. A month later she returned, by arrangement, to be treated for purulent endometritis. The cervix was dilated, the interior of the uterus curetted, and

Churchill's iodine solution applied on cotton wool. She was discharged in three days, feeling quite well. She presented herself fifteen months afterwards, and was quite well. She had remained free from pain and discharge, and menstruated regularly. A vaginal examination revealed nothing abnormal in the pelvis.

This case is a typical example of pyosalpinx, from the spread of gonorrhœal infection along the endometrium to the tube. When once pus has collected within the tube, there is no way of escape for it but in a vicious direction, and hence the only satisfactory method of treatment is to remove it by operation. The case again illustraets the uselessness of the thermometer as a test of the presence of pus in the pelvis, a much safer criterion of which is the occurrence of repeated attacks of pelvic peritonitis.

CASE 16. *Menorrhagia and dysmenorrhœa for fourteen years; occasional treatment by pessaries, dilatation, and hot injections, with only temporary relief; right side of pelvis and retro-uterine pouch occupied by an irregular swelling, thought to be due to disease of right tube and peritoneum; abdominal section; inflammation of right tube, right broad ligament, and pelvic peritoneum; cystic disease of right ovary; left appendages apparently normal; removal of right tube and ovary; death from shock (?); autopsy; uterus and left tube full of purulent mucus.*—Lydia B—, a single woman aged 34, head nursemaid in a private family, was sent to me for my opinion under the following circumstances. Menstruation commenced at the age of fourteen, and was regular and painless until the age of nineteen, when she began to have pain before the flow, and the periods became more frequent and the loss greater. This went on for four years before she sought advice. She was then examined, and was told that she had inflammation of the womb. She was laid up at home for seven weeks, and injections of hot water were ordered. After that she attended as an out-patient at the Soho Hospital, and wore a pessary for three months.

Three years later, being no better, she saw Dr. Braxton Hicks at Guy's Hospital, who said the passage was too small, and would have to be stretched. She was an in-patient for a fortnight, when she underwent an operation the nature of which she did not know. She afterwards attended at Guy's as an out-patient, and wore a pessary for nine months. For the next four years her condition was improved, though she never felt well. In June, 1888, she became worse, and in December the pain was so severe and she was so faint and sick at each period that she again took medical advice. The passage was again declared to be too small, and was dilated on two occasions just before her periods. She was ordered four or five hours' rest every day, and hot vaginal injections. She was said to have descent of the womb, and a ring pessary was inserted, which she wore for three months. She then came under the care of Mr. Hosking, of Turner's Hill, complaining of severe pain in the lower part of the abdomen on the left side, and in the right leg. Examination *per vaginam* occasioned great pain, especially on the left side. Nothing gave relief but morphia and rest in bed. These did much good, except to the pain in the left side, but at the next period all the suffering returned, and the pain became so constant that Dr. Hosking advised that another opinion should be taken.

The abdomen was normal in appearance and on palpation. On bimanual examination *per vaginam*, the uterus was found fixed, normal in size and position. There was no depression of the lateral fornices. An irregular nodular swelling filled up and depressed the retro-uterine pouch. In connection with it a sausage-shaped mass could be traced from the right side of the uterus, twisted upon itself, and descending backwards and inwards towards the swelling in Douglas's pouch. An examination of the left side of the pelvis caused more pain than the right, but nothing abnormal was detected to account for the tenderness.

The diagnosis was suppurative salpingitis with pelvic peritonitis. Operation was advised and agreed to.

On opening the abdomen the right Fallopian tube was found thickened, and bent backwards and inwards in the direction of a mass filling up the pelvis behind the uterus, and intimately adherent to the surrounding peritoneum, which was enormously thickened. The separation of this mass was difficult, and took up much time. During the manipulations a quantity of thin fluid, of a reddish-brown colour, escaped. On bringing the mass into view it was seen to consist of the cystic right ovary (the largest cyst in which had burst), embraced by the right Fallopian tube and broad ligament, both of them many times their normal thickness. The tube was empty, and its lumen not appreciably widened. The broad ligament was soft and friable, and the ligature cut through it, necessitating a second ligature around the pedicle. The left adnexa were to all appearances normal, and were not disturbed. The operation lasted 1 hr. 10 min.

Seven hours after the operation the patient had not rallied from the shock, and on removing the dressings the pads were found so saturated with blood that it was decided to reopen the wound and search for the bleeding point. This was done, but no bleeding point was discovered. The pedicle was further secured by another ligature. The infundibulo-pelvic ligament was also transfixed and ligatured to make sure of the ovarian artery.

The patient never rallied. The legs were bandaged in flannel, ether was given subcutaneously, brandy and water and champagne were given by the mouth, and, lastly, the patient was placed in a blanket-bath, but all to no purpose, death occurring forty-seven hours after the operation. There was no vomiting throughout, but there was more or less suppression of urine from the time of the operation. The quantity drawn off on the 25th was as follows:—at 2 a.m., 4 fl. oz.; at 9 a.m., 6 fl. dr.; at 4 p.m., none; at midnight, 1 fl. dr.; and at 4 a.m. on the 26th, none. The temperature an hour after the operation was

96° ; for the next twelve hours it was from 98° to 98·6° ; then it rose 100·4°, and from that time forward varied from 100·8 to 101·6.

Dr. Hadden made a post-mortem examination forty-eight hours after death. The wound had united. The intestines were distended. The stomach contained much dark green fluid. There was no peritonitis and no blood in the peritoneal cavity. The bladder was empty. The right ureter was carefully dissected out and found intact. Kidneys healthy and pale. Lungs gorged with blood. Heart nearly empty ; firm clot in right auriculo-ventricular valve. Uterus large, some muco-pus in its cavity ; lining membrane hyperæmic. Left Fallopian tube normal in length, consistence, and general appearance. On section it was, however, found to contain thick muco-pus along its entire length.

I have described this disappointing case so fully that my comments upon it must be brief, although many points suggest themselves for remark. What was the origin of the pelvic inflammation ? Was it septic ? If so, was the infection conveyed on one of the occasions when the cervix was dilated ? Was the cystic condition of the ovary secondary to the peritonitis ? What is the lesson to be learnt from the fact that the apparently healthy left tube was found after death to be full of pus ? Is it that where one tube is found manifestly diseased both tubes should be removed ? What was the cause of death ? If it was shock, why was the shock so profound ? Was the reopening of the abdomen in any way accountable for the fatal result, and was it justified ? These are some of the questions that suggest themselves—questions, it seems to me, more easily asked than answered.

Cases 17 to 21 have been published very fully elsewhere : Cases 17, 18, and 20 in the 'St. Thomas's Hospital Reports,' vol. xix ; Cases 19 and 21 in the 'Brit. Med. Journ.,' December 27th, 1890.

CERTAIN CASES OF PELVIC PERITONITIS. 49

CASE 22. *Sudden pain in left iliac region six months after an attack of phlegmasia dolens in the left leg; soft non-fluctuating swelling on left side of pelvis displacing uterus to right; patient very ill, with high temperature; abdominal section; mass situated between layers of left broad ligament, with surrounding adhesions; appendages healthy; adhesions partially separated; tumour not disturbed; abdomen closed; recovery.*—A married woman, aged 27, was admitted January 10th, 1890, looking pale and ill, and complaining of great weakness and of severe pain in the left iliac region. She had borne seven children, all her labours having been easy and natural until the last one, which took place in April, 1889. On that occasion the arm presented, and delivery took place under an anæsthetic in the Maternity Home at Battersea. About three days after delivery patient was hot and restless, and had pains all over. She rose on the tenth day, but was at once seized with pain in the left leg, and returned to bed. " White leg " supervened, and patient was laid up seven weeks in the hospital and three weeks at home. After this she felt well, though the leg ached in wet weather. Menstruation became re-established, and continued regular up to her present illness.

On January 4th, in an interval following a menstrual period, patient was suddenly seized with acute pain in the left side of the lower part of the abdomen, obliging her at once to discontinue her work and go to bed. Four days later, the pain being still present, she commenced to vomit, rejecting everything she took.

On admission the abdomen presented a normal appearance. No tumour could be seen or felt. There was some tenderness with a sense of resistance over the left iliac region. Bimanually the uterus was found anteflexed, the fundus being pushed somewhat to the right. The left fornix was depressed, the bulging having an even and regular contour. The tenderness was too great to permit of a very thorough examination, but a swelling of considerable size could be made out on the left side of the uterus, elastic but not fluctuating. The tissues around the upper portion of the cervix were swollen both in front and behind.

On the 13th January the pain had increased, especially towards the back. There was a sensation of pressure on the bowel. The patient was very ill, and the temperature varied between 100·2° and 104°.

The diagnosis being pelvic abscess, it was determined to open the abdomen the following day—if the symptoms were not relieved in the meantime. Next day there had been a slight purulent discharge from the rectum, mixed with much mucus, and the patient appeared much relieved. The operation was therefore postponed until the 17th. The temperature on the 14th varied between 101° and 104·2°, on the 15th between 99° and 101°, and on the 16th between 98° and 99·4°.

A further vaginal examination was made on the 15th. The œdematous swelling about the vaginal reflection had disappeared. Through the left fornix could be felt a large tense mass, tender to the touch, continuous with a swelling behind the uterus, pushing it forwards and to the right. The vagina was shortened on the left side, but there was no brawny condition of the roof, such as to indicate the presence of cellulitis.

On opening the abdomen the swelling was found to be situated between the layers of the left broad ligament. Its surface was even, and its consistence soft but solid. It distended the broad ligament along its whole length, displacing the uterus forwards and to the right. There was no sulcus between the uterus and the swelling, the uterus being differentiated only after inserting a sound. Posteriorly the mass was adherent to the tube, ovary, and pelvic wall, and there was some adhesive peritonitis to the right of the uterus. After separating some of the adhesions it was decided not to interfere further, it appearing probable that the swelling was a hæmatoma. The right tube and ovary were normal, and lay behind the displaced uterus. A drainage-tube was passed into the retro-uterine pouch and the abdomen closed. The tube was removed in six hours. The temperature for the first two days ranged from 99·4° to 102·6°; after that it seldom exceeded 100°.

Three weeks after the operation the mass had diminished considerably, especially at its outer part, both in height and thickness. The uterus was in the middle line. A week later the patient went home nearly well. I met her some weeks afterwards. She was very well, though still conscious of discomfort on the affected side after prolonged exertion.

It seems probable that the pus discharged from the rectum with such signal relief to the symptoms, four days after admission, was due to the bursting of a small abscess. There was no evidence of fluctuation in the tumour when exposed at the operation, and it was, therefore, not meddled with. The suddenness of onset led me to regard the effusion as a hæmatoma of the broad ligament. The peritonitis was evidently secondary.

CASE 23. *Sudden attack of pain nine weeks after last menstruation, followed by a hæmorrhagic discharge from the vagina continuing for three months, with an intercurrent attack of inflammation; elastic non-fluctuating mass behind uterus and left broad ligament; no change after a fortnight's rest; abdominal section; mass of old blood-clot enucleated; uterine appendages not disturbed; recovery.*—A married woman, aged 28, the mother of two children, was admitted January 6th, 1890. She stated that on September 20th, 1889, nine weeks after the last menstrual period, she was seized somewhat suddenly with pain in the lower part of the abdomen, of an intermittent character, with nausea and faintness. She did not think she was pregnant at the time, nor does she think so now. Two days after this attack a hæmorrhage from the vagina commenced, and this has continued almost without intermission up to three days before her admission, that is, for over three months. On the ninth day she had to lie up for what was said to be inflammation of the womb, and remained in bed for three weeks. Defecation was preceded by severe pain.

On admission she was somewhat anæmic. She had no abdominal swelling; the abdominal muscles were flaccid; there was some tenderness in the left iliac region. An oval swelling of the size of an orange was discovered, on bimanual examination, behind the uterus and the left broad ligament. The swelling was smooth and elastic, but non-fluctuating; its long axis was directed forwards and to the left; it was moveable within certain narrow limits, and could be traced as being closely connected with the left uterine appendages.

There was at this time no hæmorrhage or pain. The temperature was normal.

No change having taken place in the swelling after a fortnight's rest in bed, abdominal section was proposed and agreed to.

The operation took place on January 21st, 1890. The uterus was pushed forwards by a mass behind, which was closed in by adhesions. The adhesions having been partially separated, the mass was enucleated, two or three fluid ounces of serum escaping during the process. When removed, the mass was found to be composed of an outer wall of firm blood-clot, containing within it a quantity of soft, disintegrating blood-clot of a brown colour. No trace of organised structure could be detected. The mass measured 3 inches by 2 inches. The cavity left had a smooth internal surface, and was encircled by the broad ligaments, tubes, and ovaries, and posteriorly was bounded by the pelvic wall. The uterine appendages presented no marked lesion, and were not disturbed. The cavity was douched with hot boracic solution; a drainage-tube was inserted, and kept in for forty-eight hours.

The patient made a rapid recovery, the temperature never exceeding 100°.

Three weeks after the operation a vaginal examination was made. No swelling or other abnormal condition was detected. She went out next day, feeling quite well.

It seems to me very probable that this was a case of so-called tubal abortion. As no fœtal remains, however, were discovered, the origin of the hæmatocele must remain mere matter of conjecture.

This patient would no doubt have made a satisfactory recovery without operation. Had I diagnosed the case as one of hæmatocele, I should have certainly postponed operative interference, and tried the effect of prolonged rest. Having, however, opened the abdomen, it was obviously one's duty to remove the effused blood. This was all that was done, and the only result of the operation to the patient was that her recovery was hastened.

CASE 24. *Pelvic peritonitis with constant vomiting, following a chronic purulent vaginal discharge; abdominal section; chronic interstitial inflammation of both tubes with adhesions matting together tubes and ovaries; both tubes and both ovaries removed; persistent vomiting during convalescence with alarming prostration; recovery; re-establishment of menstruation.*—A muscular, healthy-looking girl aged 18, a lady cricketer by profession, applied for admission on March 24th, 1890, on account of severe pain in the left iliac region, which had commenced four weeks previously during a menstrual period. She was unmarried, but had been leading an irregular life since the age of sixteen. She stated that she had had a yellowish-white discharge from the vagina for two years, and that lately the discharge had become thicker and yellower. A week before admission she had a lump in the left groin.

On admission her temperature was 100·2°. The abdomen was somewhat distended, and its walls rigid. A bimanual examination revealed a tense, somewhat elongated swelling, the size of a small apple, in the left posterior quarter of the pelvis. High up behind the uterus was a small hard body, thought to be a displaced and adherent ovary. The uterus was normal in size and position.

A fortnight after admission vomiting set in, and for several days every meal was rejected. There was pain in the back and at the epigastrium. The swelling in the side of the pelvis had become more defined, and it could now be made out that the left tube was thickened and adherent, and embraced an ovary of the normal size. No swelling was detected on the right.

Abdominal section was performed on April 10th. The tube on each side was found thickened and adherent. The ovaries were healthy, but so completely involved in the adhesions that it was necessary to remove them along with the tubes. The right tube was the thicker of the two, being $\frac{1}{4}$ inch in diameter. The diameter of the left tube was equal to that of a large goose-quill. Neither tube contained pus. The mucous membrane was normal in appearance. The fimbriated extremity in both tubes was bent sharply on itself, the orifice in each case barely admitting an ordinary internal sound. There was some vomiting for the first two days. It then ceased for two days, but on April 14th it recommenced without obvious cause, and continued day by day until the patient's condition became alarming. She lost flesh, and became dark and sunken about the eyes. On April 25th the climax was reached. The resident was summoned at seven in the morning. He found the patient very ill, with a dry coated tongue and a pulse of 130, and complaining of severe abdominal pain. A few hours later the pulse was 140, the voice had changed and become hollow, and the dark rings around the eyes were very marked. She complained of a sensation in the head as of "a raging storm." She had all the appearance of impending death. From that time, however, she gradually improved. The voice resumed its natural tone, and on the 29th April the vomiting finally ceased. On the 1st May her appetite returned, and on the 14th she was able to be sent to a convalescent home.

I did not see her again until the 25th July, 1891, when she presented herself at the out-patient room, having had continuous hæmorrhage for seven weeks. She was looking well and in good condition. She had had no pelvic pain since leaving the hospital. Two months after the operation she menstruated, and had menstruated regularly up to the commencement of the hæmorrhage for which she now sought advice. She had been employed as a waitress, and had been on her feet all day. She had had no flushings of the face, but had recently been subject to fainting. On examination the uterus was normal, and no swelling could be felt on either side of the pelvis. On March 5th, 1892, she was in excellent health, and was still menstruating regularly.

I was loth to operate on so young a patient, and did so only because I believed that there was suppuration in the pelvis, and that the vomiting and rise of temperature were due to septic absorption. I was surprised not to find pus. On another point the diagnosis was defective. I had only discovered the mischief on the left side, whereas that on the right was even more marked.

The vomiting, which assumed such a dangerous form during convalescence, I am quite unable to account for, unless there was some independent affection of the stomach. The bowels acted well, and there was no albuminuria.

It is interesting to note the re-establishment of menstruation, notwithstanding the removal of the ovaries and tubes. The important point is that the pelvic pain has entirely disappeared, and that the health is completely restored.

CASE 25. *Severe attack of pelvic peritonitis, lasting four months; temporary improvement, followed by a recurrence of the inflammation, with general abdominal swelling and symptoms of septic absorption; occasional discharges of offensive pus from the bowel; ill-defined dulness and resistance on left side; abdominal section; two suppurating tubo-ovarian cysts, one on each side, that on the left situated in the abdomen, that on the right in the pelvis ; rapid pulse for four days ; pain and rise of temperature during second, third, and fourth weeks ; recovery.*—The wife of a shopkeeper at Slough was admitted to St. Thomas's Home on May 17th, 1890, under my care. She was thirty-eight years of age, and had no children. She had a miscarriage six years ago, and has had more or less pain on the left side ever since.

In August, 1889, while on a visit to Margate, she got wet. The next day she became overheated whilst dancing, and the day following was out for some time on the water. Next morning she was taken very ill with severe pain in the lower part of the abdomen. Being no better after four or five days, a doctor was called in, and she was in bed and very ill for three or four months. Twice during this illness she had a discharge from the bowel of horribly offensive pus. At the end of the period named she was sufficiently better to get up and walk about. She returned home to Slough, but in a week or two became ill again. There had been hardness and resistance in the left iliac region whilst at Margate. Now there occurred general abdominal swelling, and patient became even more sick than during her previous illness. She also had difficulty in micturition. In March, 1890, when her present medical attendant was first called in, her temperature averaged 100° in the morning and 103° in the evening. There was obstinate constipation and an irregular swelling in the abdomen, extending on the left side to the lower costal cartilages, fairly smooth and soft on the left side, lumpy about the umbilicus and dull all over on percussion. On bimanual examination the lumps could be moved *en masse* between the hands. High up behind the cervix was a uniform elastic swelling. On March 5th there occurred for the third time an offensive purulent discharge from the bowel. This gave great relief, and was followed by a fall of temperature. On March 10th the temperature became normal, and has remained so. A menstrual flow occurred in March, but not since.

On admission there was considerable swelling of the abdomen, with hardness and resistance on the left side, and a rounded prominence in the middle line. The cervix uteri was pushed upwards and forwards, the os uteri being above the upper margin of the symphysis pubis. The sound passed three inches; its point entered the rounded swelling in the middle line of the abdomen, and could be distinctly felt an inch below the umbilicus and a little to the right. A large, smooth, uniform, fluctuating swelling occupied and depressed the retro-uterine pouch. There was a distinct sulcus between the swelling and the posterior vaginal wall.

The patient was not in pain, but was extremely ill and helpless. The motions were still offensive, and the bowels did not act without assistance.

An enema was administered, and brought away large masses of hard lumpy fæces.

Abdominal section was performed on May 19th. The uterus was situated high up in the middle line immediately beneath the abdominal wall. It was enlarged, and presented on its peritoneal surface several sessile fibroids. Both Fallopian tubes were thickened and elongated, and lay stretched over the surface of large thick-walled cystic swellings. That on the left side extended from below the posterior part of the brim of the pelvis upwards to the lower costal margins; it was adherent to the omentum, to the peritoneum lining the iliac fossa, to the posterior surface of the broad ligament, to the tumour on the opposite side, and to the back of the cervix uteri. Its wall gave way during removal, and about a pint of bloodstained pus escaped. The other cyst connected with the right tube, dipped deeply behind the uterus, filling the sacral cavity and right side of the pelvis. It was of a similar character to the one on the left side. During the separation of the extremely firm adhesions to the cervix uteri and right broad ligament the wall of the cyst gave way. The tumours were removed by transfixion, ligature, and division of their pedicles, consisting of the uterine

end of the tube and the thickened broad ligament. A
large thick mass of inflamed and adherent omentum was
ligatured and removed. The operation lasted two hours;
at the end of the first hour the patient became very livid,
and remained so to the end.

The tumours proved to be suppurating ovarian cysts,
with the Fallopian tubes opening into them. The opening on the left side was large enough to admit the little
finger, that on the right was smaller. The left cyst in
its empty and collapsed state measured four inches in
diameter; the right measured six inches in its long diameter, five inches in its shorter. On the wall of the left
cyst were several daughter-cysts.

The patient eventually made a good recovery. The
pulse was rapid (over 130) for the first four days, though
the temperature was normal. There was no sickness or
abdominal distension. The bowels acted spontaneously
on the fifth day, and more freely after an enema. During the second, third, and fourth weeks the temperature
and pulse rose, and there were some dulness, pain, and
resistance in both iliac fossæ, with slight purulent discharge from the lower angle of the wound. In the fifth
week the pain became much less and the temperature and
pulse normal. There was still some purulent discharge
when she left the home.

Six months after she went home her medical attendant
wrote to me that the patient was walking about and
attending in her husband's shop. There was still some
purulent discharge from the sinus, generally very little,
but sometimes a good deal. She had menstruated four
times.

In February, 1891, she was stout and well. There was
no swelling discoverable on pelvic examination. The
sinus was still discharging, but very slightly. Menstruation still continued, sometimes every month, and sometimes at intervals of two months. Her only complaint
was of backache after exertion. I saw her again in
October, 1891; she was still looking well. The sinus was

discharging very slightly indeed. She serves in her husband's shop twelve hours a day. Menstruation has only occurred once during the past eight months.

August 16th, 1892.—Sinus closed eight months ago. Slight tendency to hernia at lower end of scar. Menstruation irregular, sometimes every month, sometimes every three months. Complains of indigestion, otherwise quite well.

It may be objected that this case would have appeared more appropriately under the head of ovariotomy than in the present series. To this I would reply that when the patient first came under my observation there was no definite abdominal tumour to be made out, and that the data necessary for arriving at a detailed diagnosis were not available. The operation was undertaken, as a matter of fact, for the relief of recurrent peritonitis, believed to be due to pelvic suppuration.

Nothing short of bold surgical treatment could, in my opinion, have saved this patient's life. The operation, as may be imagined, was difficult and prolonged, and indeed dangerous; but what was the alternative? Either things must have been left to take their course, or one might have been content with emptying and draining the suppurating cysts instead of removing them. In the former case death would almost have been inevitable; in the latter, even supposing recovery to have taken place, it would have been much more prolonged, and would almost certainly have been followed by a serious hernial protrusion at the abdominal wound.

PART II.—CASES 26 TO 50.

CASE 26. *Pelvic peritonitis with signs of tubal inflammation on both sides, and a small tense swelling on left side of pelvis pushing uterus to right of middle line; further development whilst under observation; abdominal section; both tubes thickened, adherent, and occluded; left ovary enlarged and cystic, one cyst suppurating; removal of right tube, and of left tube and left ovary; abscess at lower angle of wound on the eleventh day; recovery.*—The patient, a laundress, aged 25, married, was admitted May 5th, 1890, on account of pain in the left iliac region, hæmorrhage, loss of appetite, and general weakness. She had borne one child at full term in January, 1888, her recovery being on that occasion rapid and satisfactory. In April, 1889, she miscarried, at two months, of twins. Since then she has never felt well, but there were no definite symptoms until twelve months after, namely, three weeks before her admission. Menstruation had been regular. Her last period commenced on April 23rd, 1890; it lasted a week. A day or two later the flow recommenced and has continued up to her admission, the discharge latterly having been dark and clotted. There has been a good deal of pain in the left iliac region, and during the past few days there have been pain and difficulty in micturition, pain in the lower bowel, and constipation. There is no history of sickness or sudden pain. Patient has lost flesh; her appetite has failed, and she feels weak.

She is a light-complexioned woman, pale and anæmic.

Nothing abnormal is noticed in the appearance of the abdomen. On palpation a swelling is felt in the left iliac region, with a well-defined upper margin an inch above the level of the anterior superior spine. It extends a little beyond the middle line towards the right side. It is comparatively dull on percussion, and somewhat tender.

The uterus is fixed, slightly retroflexed and deflected to the right side, and of normal length. The left fornix is somewhat depressed by a tense, tender swelling in the left posterior quarter of the pelvis, extending from the uterus outwards to the lateral wall of the pelvis, and moveable to a very limited extent antero-posteriorly.

The temperature and pulse normal.

Urine free from albumen.

During the fortnight following admission the physical signs underwent several important changes. On May 10th there was observed a tongue-shaped, smooth, elastic swelling between the rectum and the upper part of the posterior vaginal wall, evidently due to effusion in Douglas's pouch. For several days patient suffered from vesical and rectal tenesmus, and frequently passed mucus like white of egg from the bowel. Otherwise she was feeling well and free from pain. The temperature ranged from normal to 101·2°. On May 15th the posterior swelling had become smaller and less tense; that on the left side remained as before. On the 18th both Fallopian tubes were felt thickened, their outer portions flexed and adherent behind their respective broad ligaments. The left tube lay in a plane somewhat anterior to that in which the right one was lying. The uterus was still a little to the right of the middle line, and the swelling on the left remained unaltered.

Abdominal section was proposed to the patient, but as her pain had greatly diminished she did not at first give consent. In a day or two, however, she expressed her willingness to undergo the operation.

On May 22nd an incision of three inches in length was made in the middle line of the abdomen. The uterus was found fixed to the right of the middle line, the upper part of the cervix being adherent posteriorly, obliterating Douglas's pouch. The right tube was flexed upon itself, and adherent behind the right border of the uterus. The ovary, normal, but universally adherent, was embraced by the tube. The left tube was found with difficulty. It was coiled upon itself and lay embedded in a cavity shut off by adhesions. The left ovary was enlarged to the size of a hen's egg, cystic and adherent throughout. It lay behind the left broad ligament. The left ovary and tube were separated first, the manipulations being difficult and prolonged. In bringing the parts to the surface two cysts gave way in the ovary, one containing blood-stained mucus, the other purulent fluid. The fimbriated end of the tube was occluded, the fimbriæ being indistinguishable. About $1\frac{1}{2}$ inches of the outer end of the tube was removed with the diseased ovary. The inner portion of the tube—measuring about 3 inches in length—was left, the adhesions being so firm and deeply seated as to render removal nearer the uterus almost impracticable. The right tube was now with much difficulty separated from its adhesions. The whole tube, thickened to the size of the forefinger, was removed. Its fimbriated end was occluded, and around the closed end was a mass of hard yellow material like altered lymph. The ovary being of normal size, and being bound down by a firm peritoneal band, was left undisturbed. The abdominal cavity was douched and a drainage-tube inserted. The operation lasted an hour and a half.

The patient had no untoward symptom until the seventh day, when she did not feel well, and there was some circumscribed hardness and tenderness on the left side. On the eleventh day a large quantity of thick blood-stained pus escaped from the lower angle of the wound. After this the patient was much more comfortable, and the hardness disappeared. A fortnight later she went home, the discharge having almost ceased. A stitch came away at the end of August, and the sinus then closed.

She presented herself, at my request, on September 26th, 1891, looking stout and well. She had had no pain, had menstruated regularly, and had been in perfect health ever since her discharge from the hospital.

October 22nd, 1892.—Continues well and strong, and free from pain. Menstruates regularly. Has a slight hernial protrusion in two or three places along the wound.

The portion of the left tube removed was much thickened. Its mucous membrane presented a number of minute transparent elevations (? tubercles), and in the tube wall there were several patches of softened and discoloured tissue of the size of a pea. Some flakes of lymph were found in the tube, but no pus. The right tube only contained some mucus.

This case is a typical example of a very numerous group, in which inflammation of both Fallopian tubes is associated with cystic degeneration of one or both ovaries, and in which the seat of the suppuration underlying the pelvic peritonitis is not tubal but ovarian. Very often, as in this case, only one comparatively small cyst is found in a state of suppuration, the remaining cysts containing fluid of the character usual in cystic adenoma of the ovary. Sometimes the contents of several of the cysts have become purulent. This suppuration is probably owing to the invasion of the ovary by infective micro-organisms from the neighbouring tube.

These cases seem to me to have an important bearing on the etiology of suppurating tubo-ovarian cysts, as showing that ulceration of the walls of an ovarian cyst leading to a communication between tube and ovary may commence from within as well as from without,—that is to say, may originate from suppuration within the ovary as well as from suppuration in the tube. I shall have something more to say on this subject later on.

CASE 27. *Acute pelvic inflammation following exposure to rain and cold six weeks after confinement; recovery, followed by fourteen months of apparently good health; readmission for persistent hæmorrhage; signs of chronic pelvic peritonitis, with thickening, displacement, and fixation of both tubes; removal of both tubes and both ovaries by abdominal section; loose pus-cells discovered by the microscope in the lumen of the tube, but no purulent collection visible to the unaided eye; abdominal ostia of tubes much contracted and adherent; recovery.*—A young widow, aged 23, was first admitted into Adelaide Ward in December, 1888, nine weeks after her first confinement. She had been an inmate of the surgical wards on account of breast abscess, and, happening to be discharged on a wet day, got her clothing wet through on her way home. During the night she was attacked with severe pain in the lower part of the abdomen, shooting down the thighs. The pain continued up to the time of her admission, three weeks later. On admission a hard ridge of inflammatory exudation was felt between bladder and cervix (anterior parametritis). A few days afterwards the hardness, diminishing in front, had extended to the left broad ligament, which could be felt as a hard flattened mass, moveable within certain narrow limits independently of the uterus. On January 4th the cellulitic exudation had subsided sufficiently to permit the mapping out of the Fallopian tubes, which could be felt as firm cords running outwards, one on each side, from the body of the acutely anteflexed uterus along the free border of the broad ligament. The patient rapidly improved, and left the hospital free from pain and with a normal temperature on January 19th, 1889.

After leaving the hospital she remained well and able to do her work for fourteen months. At the beginning of April, 1890, after having had a yellow vaginal discharge for a fortnight, she had to leave her work on account of persistent hæmorrhage.

On May 17th, 1890, she was readmitted to Adelaide

Ward, the hæmorrhage having then lasted for five weeks. On the 23rd she was examined under an anæsthetic. The uterus was in normal position. There was a hard irregular swelling in both posterior quarters of the pelvis, more marked on the right, where the tube could be made out distinctly as a thick cord bent backwards upon itself, and dipping down behind the uterus. On the same side there was also a softer and more circumscribed swelling, thought to be the prolapsed ovary. There was much tenderness in the situation of Douglas's pouch. Four days later there was a tense fluctuating swelling in Douglas's pouch. This gradually subsided, leaving an irregular hardness, distinctly nodulated, and the thickened and adherent right tube could again be clearly defined.

On June 6th an incision three and a half inches long was made in the middle line. On passing the fingers into the pelvis a quantity of serum escaped. Tracing the right tube from the cornu of the uterus it was found thickened, bent on itself, and adherent behind the broad ligament and the uterus, enclosing within its fold the ovary, which was enlarged to the size of a pigeon's egg, and contained several cysts, one of which was filled with pus. The tube and ovary were separated and removed. The left tube and ovary were universally adherent, though apparently themselves unaltered. These were also separated and removed. There were still some thickened irregular masses at the bottom of the retro-uterine pouch, but as it seemed certain these were only portions of thickened omentum they were not disturbed. The abdomen was irrigated with hot solution of boracic acid, and then cleansed by sponging. A drainage-tube was inserted and the wound closed. The operation lasted an hour and twenty-five minutes.

On examination of the parts removed, the right tube was found thickened to a diameter of half an inch, the wall, on section, measuring from three sixteenths to a quarter of an inch in thickness. The external surface was covered with vascular shreds of adherent peritoneum.

The mucous membrane was swollen and œdematous. The opening at the fimbriated end was contracted to the size of a mere pin-hole; the fimbriæ were thrown back and adherent. No fluid was visible in the canal. A section of the tube was examined under the microscope by Mr. Shattock, who reported small-celled infiltration, with a few loose pus-cells in the lumen of the tube. The left tube was thickened to the size of a goose-quill, denser and firmer than normal. The mucous membrane of the outermost inch was livid, swollen, and soft; the rest was normal. The fimbriated end was narrowed, but still pervious. Portions of both ovaries had been left in the pedicle on the distal side of the ligature.

The patient had a slight rise of temperature on the evening of the fifteenth day, with some abdominal pain, followed by swelling and tenderness behind and to the right of the uterus. These symptoms subsided in a few days, and on July 10th she was sent to a convalescent home feeling very well.

On April 2nd, 1891, I met the patient looking stout and well. Her complexion, which had been pale and sallow, had assumed a healthy colour. She was free from pain. There had been amenorrhœa for four or five months after the operation, since which time she had menstruated regularly.

January 7th, 1893.—Well and strong, and free from pain except at the menstrual periods, which are quite regular, but for the last six months have been painful. Married a second time two years ago. Vaginal examination reveals nothing to account for the dysmenorrhœa.

This was a case in which prolonged rest would in all probability have resulted in cure. The patient, however, being a widow, and dependent on her own exertions for a livelihood, naturally preferred a shorter and more certain method of treatment, fully appreciating and accepting the physiological consequences. I was surprised not to find a larger collection of pus in the tubes. The microscope, however, proved that it was there, though in small quantity. The re-establishment of menstruation was probably due to a portion of the ovary having been unavoidably left in the pedicle, the ligature below the portion so left not having completely destroyed its functions.

CASE 28. *Recurrent pelvic peritonitis; constant pain in back and lower part of abdomen; uterus elevated and displaced forwards; large tense swelling on left, hard irregular mass on right; abdominal section; purulent salpingitis on right side with suppurating intra-peritoneal hæmatocele; inflamed and adherent intestine in left posterior quarter of pelvis; left tube and ovary not found; recovery.*—A married woman, aged 34, residing at Streatham, was admitted June 23rd, 1890, complaining of constant pain in the lower part of the abdomen and in the back.

She was married at twenty-one, and had had four children, the last one five years ago. Two days after the birth of her first child she had an attack of inflammation, but she was able to be up on the tenth day, and had no further trouble. About two years ago she had another attack of internal inflammation, supposed to be due to a chill during menstruation. She was poulticed and syringed, and recovered in a few days. She was well up to three weeks ago, when she was seized, a fortnight after a period, with aching pains in her limbs; these disappeared, leaving, however, a constant pain in the lower part of the abdomen, especially on the right side, and in the back. No swelling had been noticed.

On admission she had the appearance of a pale but otherwise healthy woman. The thoracic viscera were normal, the urine healthy. The abdomen was somewhat distended. Resonance was somewhat impaired over the lower half of the hypogastric region.

On examination under ether, June 24th, the fundus uteri was found elevated to three quarters of an inch below the umbilicus. The mobility of the uterus was impaired, its cavity not enlarged. In the left iliac region was a tense cystic swelling the size of a large orange, moving with the uterus and evidently connected with it by adhesions or otherwise. It extended to within three quarters of an inch of the umbilicus. Fluctuation was distinctly made out bimanually. On the right side, high up by the side of the uterus and adherent to it, was a hard, irregular swelling of the size of a Tangerine orange, giving the impression of tube and ovary involved in a mass of adhesions.

A few days later the mass on the right side had become less distinct; that on the left remained the same.

The temperature on admission was 100·4°; afterwards it varied from normal to 99·6°.

Abdominal section was performed July 3rd. The omentum roofed in the contents of the pelvis, which were densely matted together and difficult to distinguish. The omentum having been separated and pushed aside, the enlarged and thickened right Fallopian tube was found deeply situated at the back of the pelvis and adherent on all sides. The adjoining ovary was likewise embedded in adhesions, but in other respects it was normal. Both were separated and removed. During the separation there escaped from amongst the adhesions a quantity of grumous fluid, consisting of altered blood mixed with pus. The left side was now explored. At length a thick-walled tube was discovered dipping down into the left posterior quarter of the pelvis and firmly adherent. This was separated and brought into view, when it was recognised by the appendices epiploicæ to be a coil of large intestine, inflamed, thickened, prolapsed, and adherent. The search for the left Fallopian tube was thereupon resumed, but neither it nor the left ovary were discovered. The pelvis was irrigated with hot boracic acid solution, a glass drainage-tube inserted, and the abdomen closed.

The portion of the right tube removed was three inches in length. The fimbriated end was open, and had a diameter of one third of an inch. The tube was thickened and inflamed. On section it was found to contain a few drops of pus. Its external surface presented a thickened, indurated, ragged patch, of dark colour, three quarters of an inch in length, which gave the impression of having formed part of the wall of an intra-peritoneal abscess. About two thirds of the normal ovary had been removed with the tube.

The patient had more pain than is usual during the first few days, but made a good recovery. The bowels were opened by enema on the fifth day. The temperature on the second day rose to 100·4°; after that it never reached 100°.

On vaginal examination, July 18th, a mass was felt on the left side depressing the vaginal roof; nothing abnormal on right side or behind the uterus. The patient was up the following day, and left the hospital well a month after the operation.

Two or three points in this case call for remark. In the first place, the association of salpingitis with an intra-peritoneal hæmatocele, an association observed in several other cases in this series, seems to point to a causal connection between the two conditions. Where, as in this instance, the salpingitis is purulent, the fimbriated end remaining patent, the hæmatocele almost inevitably undergoes suppuration, forming one variety of pelvic abscess. Another feature in the case worth noting is the difficulty that arose from an inflamed, prolapsed, and adherent coil of intestine simulating an inflamed Fallopian tube. It is next to impossible sometimes to distinguish, by touch alone, the one from the other, and even when sufficiently separated to be brought into view it is not always easy at first sight to say with certainty whether it is inflamed bowel or inflamed tube that lies before one. It is only by carefully tracing the tube to its uterine end, or, in the case of intestine, by noting appendices epiploicæ upon it, or by tracing it beyond the inflamed portion and finding it continuous with healthy bowel, that the diagnosis can be established. One of the chief risks of the operation for the removal of diseased tubes consists in this liability to mistake intestine for Fallopian tube. There can be no doubt that the smooth, tense swelling which was felt before the operation on the left side of the pelvis, and which was still perceptible a fortnight after the operation, was inflamed intestine, prolapsed and adherent. Now that all the pus has been removed the inflammation of the prolapsed bowel will gradually subside. If an opportunity should occur of examining the patient again I shall expect, therefore, to find the swelling much smaller in size and softer in consistence.

CASE 29. *Recurrent pelvic peritonitis following an attack of general peritonitis seven years ago at the age of twenty; attacks more frequent during last two years; dragging pain in right iliac region after the least exertion, necessitating the life of a chronic invalid; uterus fixed; hard irregular mass in each posterior quarter of pelvis, more marked on right; abdominal section; contents of pelvis densely matted; right tube distended by a mass of soft tuberculous material, its walls softened and marked by scars of old ulcers; left tube enlarged, thickened, and empty; mucous membrane granular; mass of soft tuberculous matter in left side of pelvis; lengthy operation, severe shock; slow convalescence; copious escape of pus from rectum seven weeks after operation; recovery.—*
A single lady, aged 27, had for seven years been subject to attacks of localised peritonitis in the pelvis, chiefly on the right side, following an attack of general peritonitis at the age of twenty, when she was confined to bed for six weeks. The localised attacks had been more frequent during the past two years. They usually occurred in connection with a menstrual period, and were always ushered in with acute pain and were attended with fever. During the whole time there had been dragging pain in the right iliac region after the least exertion. About two years ago there was a swelling in the right iliac region, which afterwards disappeared. For the last five or six

weeks she had been much in bed; before that she had been in the habit of rising at ten and retiring to bed at nine. She had lost a stone in weight during the last two years, but retained a good colour, and ate and slept well.

Her mother and a maternal uncle had died of phthisis. The patient herself had always been delicate. At the age of fourteen she had an illness, said to be due to some disease of the liver; this illness was followed by hysteria. She had had two attacks of pleurisy. The first menstruation occurred at the age of twenty (after the attack of general peritonitis above alluded to); since then she had menstruated regularly.

There was no unusual appearance about the abdomen. Vaginal examination, rendered difficult by the virginal condition of the orifice, showed fixation of the uterus with a hard irregular mass in the right posterior quarter of the pelvis, and a similar but less defined mass in the left posterior quarter. The vaginal roof was not depressed.

It being evident that there was chronic disease of the uterine appendages of both sides, with much matting of the parts, and probably with suppuration, abdominal section was suggested and agreed to.

The operation was performed July 10th, 1890. There were such extensive adhesions of intestine and omentum to the abdominal wall and to the anterior surface of the pelvic viscera that a long time was occupied in obtaining access to the pelvis. The contents of the pelvis were densely matted together. The right side was first dealt with. During the separation of the densely adherent tube and ovary on that side, the finger passed into a mass of caseous material, which proved to be in the interior of the Fallopian tube. The wall of the tube at this part was so soft that it gave way in its entire circumference, separating the tube into two distinct portions, an outer dilated portion and an inner portion. The outer portion was carefully separated from its deeply seated adhesions and removed. The torn end presented the appearance of an old abscess-cavity, filled with caseous material. The fimbriated end was closed. The ovary, normal in size and entirely embedded in adhesions, was then shelled out and brought into view. The broad ligament was then transfixed beneath the ovary, and the ovary and uterine end of the torn tube were ligatured and removed. On the left side the condition of the parts was extremely puzzling, so much so that I was sorely tempted to abandon the attempt to deal with it. Eventually, however, the ovary was discovered embedded in adhesions, and then the greatly elongated and thickened Fallopian tube. In separating the latter another collection of caseous material was entered by the finger. This mass was outside the tube, in a cavity formed by peritoneal adhesions. On separating the densely adherent fimbriated end of the tube there was a smart hæmorrhage. After a careful examination of the tube, to make sure it was not an inflamed coil of intestine, which it much resembled, the greatly thickened broad ligament was transfixed in the usual manner, and the tube and ovary were removed. The pelvic cavity was then douched with hot water, and afterwards well sponged. A glass drainage-tube was inserted and the abdominal wound closed. The operation lasted nearly three hours. At its close the patient was very cold, and was suffering severely from shock. Half an hour afterwards a subcutaneous injection of $\frac{1}{4}$ gr. of morphia was administered. The patient slept a little, and the surface gradually became warmer. At 10.30 p.m. the dressings were changed and the urine drawn off by catheter. The slight movement involved in the readjustment of the binder induced vomiting. The pulse was very rapid and feeble. During the night and up to 6.30 on the following day there was occasional vomiting. At 6 p.m. morphia was given subcutaneously, after which she slept for two hours. At 10.30 p.m. the condition had decidedly improved; the pulse was stronger and less frequent (128); the patient was in good spirits and begged for nourishment. On the morning of the third day she was bright and talkative, and interested in her future. An india-rubber tube was substituted for the glass one. At noon, urine was passed naturally, and at 3 p.m. flatus escaped *per anum*. The india-rubber tube was only kept in for twenty-four hours. The bowels acted slightly on the fifth day and copiously on the seventh. The stitches were removed on the eighth day. On August 6th, a month after the operation, the patient left for Eastbourne. Her temperature had never reached 100° up to that time, nor had she complained of any pain or discomfort. Ten days after her arrival at Eastbourne she had a rigor with slight rise of temperature, and some local tenderness in the left iliac region. A fortnight later (September 2nd) a little pus was observed in the stools, and next day 10 or 12 fl. oz. of pus passed *per rectum*. On September 19th the temperature still remained high, and more or less pus continued to be passed from the bowel every day. On October 15th the patient was eating well and was very comfortable, but the temperature, normal in the morning, rose every evening to 101°, and night-sweats were reported to be constant. On October 19th Mr. Ewart, of Eastbourne, made a vaginal examination. The uterus was retroverted and fixed. There was hardness low down in the recto-vaginal septum and all round the rectum.

After this she slowly improved, and in January, 1891, she was able to drive about in a sledge, and had lost almost all the aching pain in the pelvis which she used to suffer after walking.

I saw her at my rooms on September 28th, 1891— fourteen months after the operation. She was then looking well and cheerful, but she had not yet reached her normal weight. She could walk a mile without discomfort. There was a little purulent discharge from the rectum almost every day. Occasionally there was a darker discharge with pain and rise of temperature; for example, a fortnight before the interview the temperature, for four or five days, was 100° to 102°, being usually normal or subnormal. The catamenia have not been reestablished, and the frequent flushings of the face seem to indicate that menstruation has ceased. She is free from pain except under the occasional circumstances above noted.

In July, 1892, two years after the operation, she wrote to tell me that the discharge only appeared about once or twice in six weeks, and was then very slight. On November 17th, 1892, I saw her. She was looking and feeling very well, and had had no pain since the spring. She was leading a busy and active life, and thoroughly enjoying it. The discharge from the bowel ceased from June to September. Since then there had been a very little discharge on three occasions.

Further note on the condition of the parts removed.— Both ovaries normal. Right tube dilated, two inches in length; internal surface irregularly puckered, showing evidence of old cicatrices; rugæ obliterated; contents a yellow, putty-like substance; a portion of fimbriated end torn off and found adherent to the ovary; proximal end ragged, irregular, and softened, this condition extending to all the coats. Left tube enlarged, thickened, elongated, and occluded at its outer end. Lumen empty. Mucous membrane thickened and congested; surface granular, not ulcerated. A section submitted to microscopical examination showed no evidence of tubercle.

Although this patient was in a rank of life that enabled her to have every comfort and to take an unlimited amount of rest, the increasing frequency and severity of the recurrent attacks of local peritonitis seemed to point to the desirability of operative interference, an interference justified by the result. The tubercular disease, though local and inactive, was a continual source of irritation, and, even if it had not eventually produced general infection, would almost certainly have condemned the patient to a life of chronic invalidism for a long time to come. The formation of a pelvic abscess nearly six weeks after the operation was wholly unexpected, and proved a serious hindrance to recovery. It was well that the abscess discharged itself quickly by the bowel, or the consequences might have been still more serious. Considering the serious nature of the operation I do not think the patient's present condition can be regarded as otherwise than highly satisfactory. Her discomforts are slight, and she is able to move about and enjoy life to an extent that she had not been able to do for several years.

CASE 30. *Severe pain in left iliac region, back, and left thigh of four days' duration, with history of a similar attack three months previously after a fall; disappearance of pain twenty-four hours after admission; dense, irregular mass in each posterior quarter of pelvis; swollen and tortuous tube traceable from uterus on each side into the swelling; abdominal section; both tubes irregularly dilated, occluded, and full of pus; walls thickened and deeply ulcerated; no microscopic or other evidence of tubercle; uninterrupted recovery; patient well and strong when seen fifteen months afterwards.*—A young married woman, aged 24, the mother of three children, was admitted July 12th, 1890, on account of severe pain in the back and in the left iliac region, extending down the thighs and causing difficulty in walking. The symptoms had come on suddenly four days previously. She had once before had similar pain, viz. after a fall on the left side during a menstrual period three months ago.

There was no history of phthisis in the family. Her labours had all been easy, and recoveries rapid and satisfactory. Her youngest child was born fourteen months ago. She had had a yellow vaginal discharge for some years; it commenced, in fact, before she was married, and has never caused her inconvenience.

She was thin and anæmic, but very cheerful. The urine was clear and contained no albumen. Her temperature varied from 98·8° to 102·2°.

On vaginal examination the posterior fornix was found

depressed, and there was increased resistance in both lateral fornices. From the sides of the uterus a dense mass could be felt passing out towards the lateral walls of the pelvis, more marked on the right side. A portion of the swelling on each side can be felt as a tortuous and thickened tube traceable into the posterior fornix.

The patient had no pain after being in the hospital twenty-four hours, but as it appeared certain the case was one of chronic purulent salpingitis, abdominal section was proposed. The patient assented, and the operation took place July 21st, 1890.

The right uterine appendages were densely adherent in the right posterior quarter of the pelvis. During their separation some thick yellow pus escaped from a very small opening in the tube-wall at a point near the distal end, where the tube was distended and its wall thin. The broad ligament, thickened by cellulitis, having been transfixed and ligatured below the ovary, and a second ligature having been placed around the tube to prevent escape of its purulent contents after its division, the tube and ovary were removed. The left tube was now examined and found in a similar condition; it was accordingly separated, and, with its adjacent ovary, ligatured and removed. The abdominal cavity was well flushed with hot boracic solution, and a glass drainage-tube inserted before closing the abdomen. The pouch of Douglas being partially obliterated by adhesions, an unusually short tube (3¾ inches long) was used.

The patient's temperature rose on the evening of the second day to 101·4°, and on the evening of the third day to 100·6°. After that it never reached 100°, and from the eighteenth day was normal. She left the hospital on August 12th, free from pain and well. In response to my request she came to see me at the hospital on the 6th November, 1891, nearly sixteen months after the operation; she looked so stout and well as to be scarcely recognisable. She assured me she had been entirely free from pain from the time she left the hospital, and, in fact, had become quite stout, and was enjoying better health than she had done for several years. She has a menstrual period of normal character about every two months. There had been no symptoms of an approaching menopause.

Description of the parts removed.—Right tube, 3½ inches long, enlarged and thickened, fimbriated end closed, dilated in two places, viz. at the free end, where the diameter is 1¼ inches, and at a distance of half an inch from the uterine end, where the diameter is rather less. The larger of these dilatations is dusky red in colour and congested; the smaller has so thin a wall that the yellow colour of the pus within shows through it distinctly. The peritoneal covering of the tube is much thickened and covered with shreds of tissue, the remains of adhesions. In one or two places the adherent surfaces shows a parchment induration. At the upper and posterior border there is a portion of adherent omentum, ligatured and divided during the operation. On the anterior surface of the dilated end there are two small perforations, the peritoneum surrounding these being black. The walls of the tube are ¼ inch thick. No tubercles are visible. The mucous membrane is much swollen and ulcerated in irregular patches, especially in the dilated portions, which contains creamy pus mixed with mucus, and without odour. The floors of the ulcers are pitted and shaggy, with flakes of breaking-down tissue hanging from them. The perforation at the distal end communicates with the abscess-cavity.

The left tube is more convoluted than the right. It is irregularly dilated, the main dilatations being three in number, the largest of which is at the distal end, the smallest near the uterine. The peritoneal covering is thick, and shreddy from torn adhesions; beneath it is a small serous cyst. The wall of the tube is so thin at the dilated portions that the yellow colour of the pus inside is clearly shown in striking contrast to the dusky red colour of the rest of the tube. The mucous

FIG. 1.

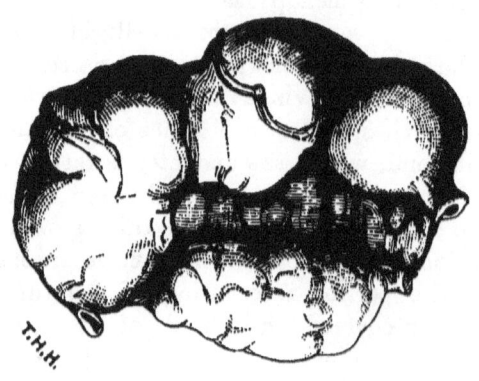

FIG. 2.

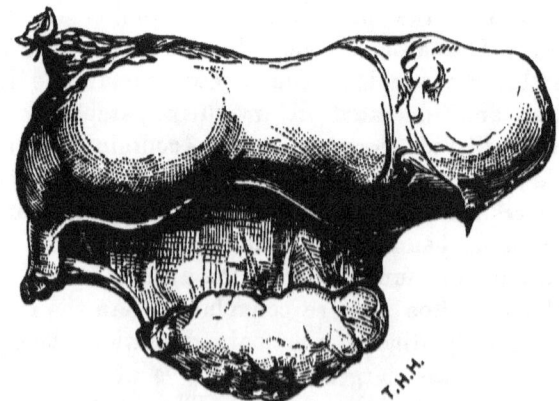

Double pyosalpinx; natural size. The drawing shows the irregular dilatations of the tubes. The ovaries are normal. (Case 30.)

membrane is generally thickened; each dilated portion is separate from the rest, and contains creamy pus without odour. The wall of the tube has, in the case of the smallest of the three abscesses, been destroyed by ulceration to such an extent that only the peritoneal coat remains. The characters of the ulcers are the same as in the right tube.

The ovaries are normal in size and appearance; they are full and pulpy, and contain a few small cysts. In the left is a recent corpus luteum.

A noticeable feature in this case was the short duration of acute symptoms, there having been but two attacks of pain, each lasting only a few days, and separated from each other by an interval of three months. Yet the condition of the tubes showed that the disease was of long standing, and that their removal was only effected just in time to avoid rupture, with escape of the purulent contents into the peritoneal cavity. The ulceration had extended down to the peritoneal coat, which itself was on the point of rupture in at least two places. This case is a sufficient answer to those who advocate a preliminary trial of palliative measures in all cases indiscriminately. A delay of even a few days would have exposed this patient to a very serious risk.

The nature of the infection seems doubtful. There is nothing in the history that points definitely either to septic or gonorrhœal infection. I suspected from the nature of the ulceration, that the disease would prove to be tubercular, but my friend Dr. W. S. A. Griffith, who very kindly removed a portion from the middle of one tube for examination under the microscope, assured me that he could discover no evidence of tubercle, although he examined several sections. The case must, then, for the present remain unclassified.

The result of the operation was, and continues to be, all that one could desire.

CASE 31. *Pelvic hæmatocele simulating cystic tumour;*

operation averted by the unexpected diminution in the size of the swelling; rapid disappearance of the tumour; exploratory abdominal section four months later, on account of persistent disablement and pelvic pain; remains of hæmatocele found, and matting of contents of pelvis; no appreciable lesion of the uterine appendages; right ovary separated and removed; right tube separated, but not removed; left appendages undisturbed; recovery; two years afterwards in excellent health and free from pain.— A married woman, aged 23, was sent from Scarborough on the 1st February, 1890, to be operated upon for an ovarian tumour. There was a fluctuating swelling in the abdomen, centrally situated, reaching upwards to the level of the umbilicus, and dipping down into the pelvis, causing in the latter situation a large bulging tumour behind the upper part of the vagina. The uterus was pushed upwards, forwards, and to the left. Menstruation had been regular. The swelling had been first noticed four months previously, being then, according to the patient's account, the size of a walnut. Three weeks before admission, at the commencement of a menstrual period, patient was seized with severe pain, which continued for the three days of the period; since that time the abdomen had been much larger than it was before. The patient had been married three years, but had not become pregnant.

I saw no reason to doubt the diagnosis of the medical attendant, except that I regarded the tumour as being a broad ligament cyst rather than an ovarian. It happened that there were several more urgent cases needing operation just at that time, and that some delay occurred in consequence. On the 17th of February, a little more than a fortnight after admission, the abdomen was observed to be decidedly less prominent and the tumour less tense. Fresh measurements were thereupon taken, and it was found that they had diminished in all directions. The upper limit of dulness, which had been $6\frac{1}{2}$ inches above the pubes, was now only 4. The distance between the pubes and the umbilicus had become reduced from 8 inches to 6½ inches, and that between the umbilicus and the anterior superior spine of each ilium from 6 inches to a little over 5 inches. The swelling, felt *per vaginam*, was also smaller and less tense, and the cervix uteri was no longer squeezed against the left pubic ramus. In short, it was evident that the swelling was a hæmatocele, and that it was disappearing. The process of absorption went on, as it usually does when it once begins, with amazing rapidity. On February 25th no tumour could be felt on abdominal examination. High up behind the cervix uteri, which was now in its normal position, could be felt bimanually a flaccid, circumscribed collection of fluid, causing little or no depression of the vaginal roof. On the 8th March the swelling was still diminishing. On the 18th it conveyed an impression very much like that given by a distended tube, and on the 25th this character was still more marked. The patient returned to Scarborough on the 28th, and was desired to present herself for examination in three months.

She came up to London again at the end of July, and was readmitted to the hospital on the 2nd of August, 1890. She had not been able to do much work during the four months she had been at home, on account of weakness and backache. Occasionally she had had pain in the left iliac region, and most of her time had been spent on the couch. She was anxious that something should be done for her if possible. There was a hard irregular swelling in the left posterior quarter of the pelvis. An exploratory incision having been determined upon, the operation was performed on August 4th. Both tubes and both ovaries were universally adherent in the posterior part of the pelvis. The tubes were not dilated or appreciably thickened. The right tube was separated as far as possible; as it appeared healthy, it was not removed. The right ovary was also shelled out from its adhesions. Its external surface was so thickened and ragged that it was thought wise to remove it. The adhesions on the left side were extremely dense, and as there did not appear to be any disease of the appendages on that side, they were not disturbed. There was a large cavity behind the uterus, with ragged walls. In this cavity were several fragments of old blood-clot. A glass drainage-tube was inserted and the abdomen closed.

The patient recovered satisfactorily, and left the hospital on August 23rd. Her general condition was improved, and she was free from pain. The uterus was fairly moveable. Some hard, irregular thickening could be felt above the left vaginal fornix, none above the right. There was slight tenderness in the situation of Douglas's pouch.

I wrote to her medical attendant at Scarborough for news of her in October, 1891. His reply, dated October 12th, stated that he had called on the patient that day. She looked very well, and expressed herself as being better than for years past. She was able to do her work, had no pain or bearing down or backache. Menstruation was regular, and, though rather scanty, was painless.

July 15th, 1892.—Presented herself at the hospital in excellent health. Has no pain, menstruates regularly, and is able to do all her housework and attend to a small business as well. On examination no swelling on right side of pelvis; left appendages adherent, otherwise normal.

This patient had lost a brother from consumption at the age of twelve. She herself had had her knee excised by Mr. Croft when she was ten years old, and, shortly after leaving the hospital on that occasion, had an attack of inflammation of the bowels. It seems not unlikely that her sterility was due to damage done to the uterine appendages by that attack. As to the cause of the hæmatocele, which I so nearly operated upon under the impression that it was a cystic tumour, the subsequent abdominal section shed no light. I think it not improbable, however, that it originated in rupture of an early extra-uterine gestation sac, or perhaps in a so-called tubal abortion. The operation was simply exploratory in its intention. I thought the persistent pain and disablement might be due to a hæmatosalpinx or some other disease of the appendages. Nothing of the kind was detected, and the patient would no doubt have recovered just as well without any operation.

The next case is one of erroneous diagnosis. I opened the abdomen expecting to find an inflamed and adherent tube and an adherent ovary lying beneath the body of a retroflexed and adherent uterus, instead of which I found no adhesions at all—nothing, in fact, but a normal ovary prolapsed beneath a retroflexed uterus.

CASE 32. *Continuous pelvic pain and dysmenorrhœa; irreducible retroflexion of uterus, with prolapsed ovary beneath it in Douglas's pouch; hard swelling, thought to be the inflamed and adherent tube, immediately above the ovary; abdominal section; retroflexed uterus; body incarcerated in hollow of sacrum from fibroid enlargement; tube and ovary normal, the latter prolapsed; no adhesions; reduction of the displacement; recovery, with complete relief of symptoms.*—A woman, aged 40, applied at the out-patient room on account of constant pain in the pelvis and dysmenorrhœa of six months' standing. She had been married since the age of eighteen, had borne one child a year after marriage, and had not been pregnant since. On vaginal examination the uterus was found retroflexed and fixed. Beneath the retroflexed body, at the bottom of Douglas's pouch, was a small soft body, thought to be the prolapsed and adherent ovary, and between the two a hard, fixed, irregular swelling, thought to be the inflamed and adherent tube. The patient was advised to come up for operation if the pain and disablement continued.

A few weeks later she begged to be admitted. She was taken into the hospital on August 4th, 1890, and the abdomen was opened the following day by an incision 2½ inches long. The retroflexed body of the uterus was found tightly incarcerated beneath the sacral promontory. On lifting it up it was seen to be enlarged and distorted by fibroids. Deep down in Douglas's pouch lay the prolapsed right ovary. Both it and the tube were perfectly normal. There were no adhesions of any kind. The ovary having also been lifted up into its proper position, a Hodge's pessary was introduced *per vaginam*, to prevent a recurrence of the uterine displacement. Until this was done the uterus showed a tendency to fall back the moment it was left unsupported. After the introduction of the Hodge it remained in its normal position. The abdominal wound was then closed. No ill effects followed the operation, and the patient went home on August 29th with the uterus in its normal position, and still wearing the pessary. She lost all her uncomfortable symptoms from that time. The pessary continued to be worn until April 18th, 1891, when it was finally removed. The uterus had remained in its normal position the whole time (nearly nine months), and the patient's health had been excellent.

I have included this case because, although the uterine appendages proved to be healthy and non-adherent, they were thought to be diseased, and the object of the operation was to remove them. The hard, fixed swelling above the prolapsed ovary, thought to be the inflamed tube, was one of several fibroids projecting from the body of the uterus. The operation furnished the only opportunity that has ever occurred to me of observing the effect of a Hodge's pessary from inside the pelvis. It raised the vaginal roof with its peritoneal covering into a distinct fold, and so far confirmed the theory that it acts by elongating the posterior cul-de-sac, thereby drawing the cervix upwards and backwards into its normal position.

CASE 33. *Small cystic ovary known for several years to be prolapsed in Douglas's pouch; gradual development in size with slight pain; sudden attack of intense pelvic peritonitis with formation of an abdominal swelling continuous with that in pelvis; subsidence and subsequent recurrence of acute symptoms; abdominal section; matting of pelvic viscera; suppurating cyst of left ovary with fœtid contents, communicating by a direct opening with inflamed left tube; right tube inflamed and adherent; diseased parts removed; death on fifth day.*—A married lady aged 31, who had formerly been under my care at Manchester, was sent up to me for operation on August 30th, 1890, under the following circumstances.

She had been married eleven years, and had never been pregnant. She first consulted me in February, 1887, having then been under medical treatment for twelve months on account of dyspareunia. At that time the only thing discovered on examination was a small, tender, cystic tumour in Douglas's pouch, which I thought was the left ovary enlarged and adherent. The uterus, normal in size, was displaced a little to the right and freely moveable. There were frequent attacks of neuralgia of the head and face; otherwise the general health was good. A year later the left (prolapsed) ovary was still very tender; the right ovary, now felt for the first time, was also tender and slightly enlarged. On December 29th, 1889, the patient came up to London to consult me. The condition then was as follows :—Left side of pelvis occupied by a tender irregular mass, partly cystic, pushing uterus over to right. Uterus normal in length and moveable. The question of operation was discussed, and it was arranged that she should see me a little later, when she had quite recovered from a recent attack of bronchitis. On August 1st, 1890, I heard that she was very ill. She had menstruated normally from July 7th to 15th, and a week afterwards had been sitting reading out of doors, and feeling perfectly well, when, on rising to go home, she found that every time she put her foot to the ground an extremely severe pain struck up into the lower part of the abdomen. She reached home with difficulty. Soon after arriving at her house she had a severe rigor with chattering of the teeth. Dr. Donald saw her the same evening, and diagnosed pelvic peritonitis. He found the left ovary enlarged and the left appendages generally matted. She improved rapidly, and in three or four days her pulse and temperature were normal, and she had no pain. On July 31st she had permission to get up, but was unable to do so on account of a return of the pain. The following day Dr. Donald discovered a large soft swelling in the left broad ligament pushing the uterus to the right. The only symptoms were backache and flatulence. The swelling was thought to be a hæmatocele. During the month of August the patient made little progress. The pulse was uniformly rapid, the temperature normal in the morning and between 100° and 101° in the evening. On the 30th August she came to London with a view to operation. Although she had been five weeks in bed she was able to walk with an ease that surprised her. She had no pain. There was a fluctuating swelling causing some prominence above the pubes, centrally situated, dull on percussion, measuring 4¼ inches transversely and reaching to within 2¼ inches of the umbilicus. Bimanually the swelling was felt to occupy the left side of the pelvis; the uterus was fixed, and lay in front and to the right.

Abdominal section was performed on September 1st. The omentum covered the pelvic viscera completely. After separating it and turning it aside, the pelvis was found to be occupied by a large mass of matted viscera, consisting of uterus, both Fallopian tubes greatly thickened and enlarged, and on the left a large thick-walled cyst. The right tube and ovary were traced out first and separated from their adhesions. The ovary was slightly larger than normal, of firm consistence, and universally adherent. The tube was thickened, dilated, and completely occluded at its distal extremity. It measured 4 inches in length, and was coiled round and adherent posteriorly. During the separation a quantity of foul-smelling pus welled up, which was afterwards found to have escaped from a rent in the cyst of the left ovary, to which the right tube had been adherent. The right tube and ovary were ligatured and removed. The broad ligament was much thickened from chronic inflammation, but was not friable. The parts on the left side were then dealt with. The tube was much thickened and elongated, and stretched out over the cystic tumour. The fimbriated end was dug out from the deepest part of Douglas's pouch with some difficulty. The thick-walled cyst, now empty, was then shelled out, and found to be an inflamed suppurating cyst of the left ovary, 4 inches in diameter. There were two small rents in its wall, and one large one. It was found that this last had been caused by tearing away the fimbriated end of the tube, which closely fitted the aperture, having formed part of the cyst-wall, and opened directly into the cyst. The left broad ligament was greatly thickened, but formed a satisfactory pedicle. The tube and cyst were removed. A coil of intestine had participated in the inflammatory process, its walls being as thick and firm as those of the inflamed tubes. The operation lasted one hour and forty minutes. The shock was alarmingly severe, and in fear lest the patient should die on the table, I did not stay to irrigate the peritoneum, but having inserted a drainage-tube and cleansed the peritoneum as well as I could, closed the wound and put her back to bed.

After an hour or two she rallied, and for the first day or two I thought the was going to recover. On the third day, however, she became very ill and restless, and on the fifth day she died. There was no post-mortem examination.

This case made a strong impression on my mind. It shows very strikingly the futility of expectant and palliative treatment where there is obvious disease of the appendages, even though the symptoms may at first be comparatively insignificant. The probability is that there was incipient and unsuspected tubal disease, in addition to the small cystic ovary, when the patient was first under treatment. I was not competent at that time to diagnose tubal disease in its early stages, and it is quite possible, if there were physical signs, that they were overlooked.

The course of pathological events seems to have been the following : Concurrent suppurative salpingitis and cystic disease of the ovary; pelvic peritonitis, with adhesion of both tubes to the gradually enlarging cyst; ulceration of the cyst-wall, ending in perforation and direct communication between the cyst and one of the suppurating tubes ; infection of the contents of the cyst, causing suppuration of contents and inflammation of the cyst-wall ; acute peritonitis and septicæmia. The operation, unfortunately, was too late to save the patient's life. No one, however, will, I think, hesitate to agree that it was the right treatment, and gave the patient her only chance.

CASE 34. *History of two attacks of severe abdominal pain, one eighteen months ago after missing one menstrual period, the other a month ago after missing two periods; soft irregular swelling behind and to the left of the uterus, extending upwards and forming a distinct abdominal tumour; ill-defined thickening of right broad ligament ; abdominal section ; ruptured blood-cyst of right broad ligament ; left hæmatosalpinx with intra-peritoneal hæmatocele ; removal of diseased parts and of right ovary ; recovery.*—A married woman, aged 29, was admitted into St. Thomas's Hospital August 19th, 1890, complaining of pain in the lower part of the abdomen and the back, and of a swelling in the left iliac region.

She was married at the age of eighteen, has had four children and no miscarriages. Her last child was born five years ago. After that she menstruated regularly until eighteen months ago, when, being a fortnight beyond her usual monthly period, she was seized very suddenly with an attack of pain in the lower part of the abdomen. This was followed a few hours later by a discharge like that of menstruation. She was seen at once by a doctor, who said she had an internal inflammation. She was on that occasion confined to bed for three weeks. She afterwards again menstruated normally until three months ago, when she missed two periods. On the morning of Sunday,

July 20th, that is a month ago, when the third period was about due, patient was again suddenly seized whilst at her housework with a very severe pain in the lower part of the abdomen, chiefly on the left side and in the back, compelling her to go to bed at once. Her face is said to have been pale and her features drawn. Hot flannels were applied all day. The pain passed off, but next day, whilst riding in a tramcar, it returned very violently, and she was seized with vomiting. Her husband met the car, and had to carry her most of the way home—about half a mile. She went to bed on reaching home, and the next day she sent for a doctor. On that day a very slight hæmorrhagic discharge from the vagina took place, which has continued up to her admission, the discharge never amounting to more than a stain. Soon after the commencement of the attack she noticed a swelling in the left iliac region. She had not considered herself pregnant, as she had not had her usual morning sickness.

On admission the patient was of a rather sallow complexion, but was well nourished.

The abdominal walls were flaccid; a firm and somewhat tender swelling could be felt in the lower part of the left side of the abdomen. The limit of this swelling in an upward direction was 2 inches below the level of the umbilicus; that on the left side reached as far as the middle of Poupart's ligament, while that on the right just transgressed the middle line. The swelling was dull on percussion.

The breasts were not swollen, but some secretion could be squeezed from the nipples.

A vaginal examination was made, under ether, on August 27th. There was much creamy mucus in the vagina; the mucous membrane was not discoloured. The cervix uteri was in its normal position. The body of the uterus was felt immediately beneath the abdominal wall, a little to the left of the middle line. The sound passed three inches. Behind the uterus was a soft irregular swelling, about the size of two fists, rising above the level of the fundus and on a plane posterior to it. The uterus could be moved to a slight extent upwards and downwards independently of the swelling behind it. There was a small, hard, moveable body felt above the fundus uteri, in front of the deeper swelling, and immediately beneath the abdominal wall. Between the uterus and the right lateral wall of the pelvis some thickening existed in the neighbourhood of the broad ligament.

Abdominal section was performed September 4th, 1890. The first thing seen was a thin-walled, dark-coloured tumour situated to the right of the middle line, with omentum closely adherent to it. The tumour was separated first from the omentum, and then from its deeper adhesions. On bringing it to the surface there was observed a rent on its posterior aspect, through which dark clot was protruding. There were many small clots of the same kind lying free in the peritoneal cavity. The rupture had evidently occurred before the operation. On the surface of the tumour the right Fallopian tube was stretched out. The tumour itself appeared to be a cyst of the broad ligament filled with blood-clot. The ovary was normal. The cyst, with the adjacent tube and ovary having been removed, the left side was dealt with. There was here a larger tumour situated deeply behind the uterus, and closely adherent to a coil of large intestine which had become prolapsed into the cavity of the pelvis, and was intimately adherent behind to the posterior pelvic wall. The abdominal incision was now enlarged upwards and downwards until it measured $3\frac{3}{4}$ inches. Beneath the tumour was a quantity of old and recent blood-clot encysted in the peritoneal cavity. The hard moveable nodule felt above the fundus uteri before operation was the smaller, uterine, end of a pear-shaped swelling, $2\frac{1}{2}$ inches long by $1\frac{1}{4}$ inches wide, consisting of the left Fallopian tube, containing a firm dark blood-clot. The tube was removed, the hæmatocele cleared out, and the peritoneal cavity douched with hot boracic solution. A drainage-tube was inserted and the abdominal wound closed. The operation lasted one hour and twenty minutes.

No trace of an ovum was discovered. The patient made a good recovery, and left the hospital, looking and feeling well, on the 4th of October, thirty days after the operation. There was a very small discharging sinus at the lower angle of the wound.

I saw her on January 5th, 1892, when she attended the hospital at my request. She has been well and at work without interruption since leaving the hospital. She is in robust health, with a good colour, and is still gaining flesh. She has menstruated regularly, commencing seven weeks after the operation. The wound is soundly healed.

October, 1892.—Stout and well; no pain; menstruates regularly.

The history of this case strongly suggests tubal gestation, but no positive evidence of it was obtained; and presuming the hæmatosalpinx and hæmatocele on the left side to have had such an origin, it is difficult to see what connection the ruptured blood-cyst in the right broad ligament can have had with ectopic gestation, unless, indeed, one supposes that the veins of the right broad ligament sharing the general enlargement of the pelvic veins due to pregnancy, one of them ruptured into an already existing broad ligament cyst. There may be a difference of opinion as to the propriety of including the case in the present series, but, as its nature was doubtful, it seemed to me, on the whole, the wiser course not to omit it.

CASE 35. *History of pelvic pain extending over a period of more than fifteen years; recurrent pelvic peritonitis during last seven years, with long intervals of apparently good health; small cystic swelling behind left broad ligament; irregular swelling on right side of pelvis, thought to be an inflamed and adherent Fallopian tube; abdominal section; old pelvic peritonitis; small, tense, thick-walled cyst of left ovary; left tube slightly thickened; right tube thickened to a diameter of half an inch, densely adherent, fimbriated end bound down and occluded; no evidence of suppuration; both tubes and both ovaries removed; quick recovery and subsequent freedom from pain, and improvement in general health.*—A married woman aged 51, a mangler, was admitted September 5th, 1890.

It was discovered when she was one year and nine months of age that she had so-called congenital dislocation of the hips. She married at thirty, and bore two children during the next three years, her labours, contrary to expectation, being easy and natural. Two years after the birth of her second child she began to suffer from aching in the lower part of the abdomen and down the thighs. Shortly after this she miscarried at two months. She recovered well, but a month later she had so much pain that she became an out-patient at St. Thomas's Hospital under Dr. Cory, and eventually an in-patient under Dr. Gervis. This was in 1875. In 1882 she again became an out-patient for bearing-down pain and yellow discharge, and a pessary was inserted, which, however, gave no relief. A year later, Dr. W. Duncan, acting for Dr. Cory, told her she had a small tumour that needed puncturing. Patient was admitted, but left the hospital in two months, nothing in the way of operation having been done. Belladonna was applied externally. Two days after leaving the hospital patient was seized with severe labour-like pains, rigors, and vomiting. She was readmitted for parametritis and enlarged Fallopian tube. Poultices were applied, and in three weeks she was pronounced by Dr. Gervis to be so much better as not to need operation. For five years she remained fairly well. Then she had another attack of pain with rigors and vomiting, and was sent to the Dulwich Infirmary, where she remained nine weeks, a recurrence of the severe symptoms occurring when she had been there three weeks. Nine months ago (Christmas, 1889) she missed two periods, and then had a rather profuse discharge, with pain in the left side. Three weeks later she had an ordinary period, and since then (six months ago) she has not menstruated at all. Four months ago she consulted a doctor for pain in the left side and a yellow discharge. Caustic was applied to the womb once a week for a month without benefit, and seven weeks ago she became an out-patient at St. Thomas's, when she was advised to submit to operation.

On admission she appeared in fairly good health. On vaginal examination the body of the uterus was found large, the cervix in normal position and fairly moveable. Bimanually, to the left of the uterus, a smooth, tense, elastic swelling, the size of a small orange, was discovered. It was quite separate from the uterus and fixed. Above the right fornix, on a plane posterior to the cervix, was an irregular, ill-defined swelling, thought to be an enlarged and adherent Fallopian tube. Abdominal section was performed on September 9th, 1890. Behind the left broad ligament was a tense round cyst, firmly wedged in the pelvis, but not adherent. It was with some little difficulty brought into view, and was found to be a single thick-walled cyst of $2\frac{1}{2}$ inches diameter, containing transparent fluid, of specific gravity 1005. The cyst and adjacent tube were removed. On the right side the tube, somewhat thickened, was prolapsed and adherent, the fimbriated end being very firmly bound down in Douglas's pouch. Surrounding the tube were several subperitoneal serous cysts. The separation was difficult owing to the firmness of the adhesions, but eventually the tube and normal ovary adjacent were tied off and removed. The body of the uterus was studded with a number of small subperitoneal fibroids. There was a good deal of oozing from torn adhesions, but no ligatures were required. The abdominal wound was closed in the usual way.

The portion of the right tube removed measured $2\frac{3}{4}$ inches in length and $\frac{1}{2}$ inch in breadth. The portion of the left tube removed was 2 inches long and only slightly thickened. There was no fluid in either tube.

Recovery was satisfactory and rapid; the highest recorded temperature being 99·2°. The patient left the hospital well on the 8th October. On February 28th, 1891, she presented herself looking remarkably well. She had gained flesh and had had no pelvic pain of any consequence since the operation. She had not menstruated. She subsequently developed a small hernia at the lower angle of the wound. Otherwise she remained well and free from pain.

There can be little doubt that an earlier operation would have saved this patient from years of suffering and ill-health. With regard to the nature of the salpingitis, the evidence is insufficient to show whether it was septic or gonorrhœal. The case shows how insufficient palliative treatment is to effect a permanent cure under such circumstances, and offers a strong argument in favour of early surgical interference.

The next case affords still stronger evidence on this point.

CASE 36. *Chronic ill-health for several years with intermittent attacks of purulent vaginal discharge and increasing dysmenorrhœa; acute symptoms of pelvic peritonitis after exposure to wet; swelling in left posterior quarter of pelvis diagnosed as thickened, tortuous, and adherent tube; uterus fixed; abdominal section; whole contents of pelvis matted by old adhesions; both tubes thickened, tortuous, and adherent, containing muco-purulent fluid; ovaries adherent, their outer coat thickened; appendages on both sides removed; recovery, followed by continuous improvement in health.*—A single woman, aged 34, employed as a barmaid, was admitted, October 9th, 1890, complaining of pain in the left side of the lower part of the abdomen and of slight hæmorrhage from the uterus. She was pale, thin, and careworn, and had the aspect of a person suffering from chronic illness. She stated that she had suffered from a thick, yellow, vaginal discharge many times during the past twelve years. For eight years she had lived as a married woman, but had never been pregnant. She has suffered severely from dysmenorrhœa from the commencement of menstruation, the pain beginning a week before the flow, becoming acute during the first few hours and then gradually abating. These symptoms have been increasing in severity during the last five years. For some months sexual intercourse has been impossible on account of the pain it caused. There has been no definite pain in the pelvis, however, at other than the menstrual periods until three weeks ago, when her present illness began, though the general health has been seriously failing for several years.

On September 23rd, the day after the last period ceased, she got wet whilst going to her work and again on returning home. On the afternoon of the 24th she began to suffer from severe pain in the lower part of the abdomen and had poultices applied. In spite of the pain she got up and went to her usual evening employment. Next day the pain was less severe, but a slight hæmorrhage commenced. She again went out in the evening. The following day she was obliged to remain in bed, and on the 27th the hæmorrhage became so profuse that a doctor was sent for and deemed it necessary to plug the vagina. During the following week, the hæmorrhage continued slightly and the doctor told her she was suffering from inflammation.

On admission, the temperature was 99°, the pulse 100, the tongue coated, the bowels confined. On examination *per vaginam*, the uterus was found fixed. In the left posterior quarter of the pelvis was a moderately hard mass, passing outwards from the uterus, then curving backwards and terminating in the retro-uterine pouch. This swelling was believed to be the distended and adherent Fallopian tube.

Abdominal section was proposed and agreed to. The operation was performed on October 16th, 1890. Some serous fluid escaped on opening the abdomen. The posterior part of the pelvis contained a mass of adherent viscera, consisting of the uterus, both Fallopian tubes tortuous and enlarged, the ovaries, some omentum and several coils of intestine. The omentum having been separated and a ragged portion ligatured and removed, the right tube and ovary were with difficulty separated and removed, the difficulty being greatly increased by the adhesion of the uterus posteriorly preventing its being lifted up so as to bring the parts well into view. The left appendages were then separated, with still greater difficulty, and removed. A small quantity of old blood-clot was found beneath the fimbriated end of the tube, surrounded by adhesions. Some coils of intestine, adherent to the posterior wall of the pelvis, were left undisturbed. The peritoneal coat of one of the coils of intestine was accidentally pinched beneath the ligature round the right tube. It was quickly set free without dividing the ligature, and the little wound closed by three fine silk sutures. There being a good deal of oozing a drainage-tube was left in and the abdominal wound was closed. The operation lasted two hours.

The walls of both tubes were found on section to be considerably thickened. There was some muco-purulent fluid in both, the quantity being greater in the left than in the right. There was no ulceration of the mucous membrane. The ovaries were large and succulent, their outer coat thickened and shaggy from peritoneal adhesions.

The patient made a good recovery. She passed flatus at 8 p.m. on the second day. She passed urine naturally on the second and third days, required the catheter on the fourth and fifth, and after that again passed urine voluntarily. The drainage-tube was removed in forty-eight hours. The bowels acted after an enema on the fourth, eighth, and eleventh days. The temperature on the day following the operation varied between 99·8° and 100·8°; for the next three weeks it was under 100°. There was a little ill-smelling pus found daily on the vaginal pad up to the 25th of October, i.e. during the first ten days. The stitches were removed on the ninth day. There was no abdominal distension throughout. There was some rise of temperature with abdominal pain and vomiting on November 12th, but these symptoms quickly disappeared. The patient was sent to a convalescent home on November 22nd; whilst there she gained 4½ lbs. in weight. On January 6th, 1891, she was still gaining weight and improving in colour and remained free from pain. There was nothing abnormal to be felt in the pelvis, save a little hardness high up behind the cervix.

On October 27th, 1891, having had a severe cold attended with some pelvic pain, she presented herself to ascertain whether there was anything wrong. On vaginal examination no abnormal swelling or tenderness was found; the uterus was moveable, and the posterior quarters of the pelvis were free. On March 10th, 1893, she attended at the hospital. There had been no menstruation since the operation. She had been at work uninterruptedly since February, 1891, and declared herself to be now in better health than she had been for several years before the operation.

Though there is no absolute proof that this was a case of gonorrhœal salpingitis, all the evidence is in favour of that opinion. The beneficial effect upon the health, of the removal of the diseased tubes, has seldom been more striking. One can scarcely recognise the patient as being the same person.

CASE 37. *Pelvic pain for six years; peritonitis twelve months ago; continually increasing pain since; admission chiefly on account of hæmorrhage due to a mucous polypus; removal of polypus; pelvic pain complained of, thought to be functional; development of septicæmic symptoms; mass discovered on one side of pelvis; abdominal exploration; both tubes tortuous, inflamed and adherent with muco-purulent contents; small cyst of right ovary full of fœtid pus; small intra-peritoneal abscess in Douglas's pouch; removal of both tubes and suppurating ovarian cyst; recovery followed by improved health; death a few months later from cancer of stomach.*—A single woman, aged 46, a housekeeper, was admitted into the hospital September 23rd, 1890, complaining of pain in the pelvis, especially on the right side, and of slight but continuous uterine hæmorrhage. The hæmorrhage dated from March, 1889, and the pain from an attack of peritonitis, in August, 1889, which was caused by getting wet, and which obliged her to keep her bed for several weeks. For at least five years before this, however, she appeared to have suffered more or less from pain in the right side of the pelvis and in the back, especially on walking or making any exertion. This pain has been much worse during the last three months.

The patient was a dispirited-looking woman, of dark complexion, of fairly healthy colour and in moderately good condition.

On September 23rd, a small mucous polypus of the cervix was removed by torsion.

On September 30th, an examination was made under ether. An irregular, hard, adherent mass was found high up behind and to the left side of the uterus. This was thought to be the prolapsed and adherent left tube and ovary.

On October 6th, the hæmorrhage had ceased, but the pain continued. I was disposed to think the patient magnified her sufferings, which at this time I regarded as largely of a functional character. A week later, however, it was observed that the patient was becoming thinner and weaker; the pain complained of was more severe, especially on the right side; the temperature rose a little in the evening; and the tongue had become dry, red, and glazed. It was evident, therefore, that there was some septic absorption going on, and I suggested an exploratory operation, which the patient readily agreed to.

The operation took place on October 23rd. The omentum, thickened by inflammation, roofed over, and was adherent to, the contents of the pelvis, which were themselves all densely matted together from old peritonitis. The right tube, much enlarged and universally adherent, was first separated. During the process, a quantity of dirty, brown, fœtid, purulent fluid welled up. When the appendages were brought fully into view, it was seen that this fluid had escaped from a small inflamed cyst of the right ovary owing to accidental rupture during separation. The remainder of the right ovary was dense and thickened from chronic inflammation.

The bladder was much thickened and the proximal portion of the right tube was intimately adherent to it. The connection was highly vascular but was separated without injury to the bladder. The left tube was much enlarged and thickened and universally adherent, its fimbriated end being very firmly adherent to the lower part of the posterior surface of the uterus. Both tubes were removed. There were still remaining some hard irregular masses in the left posterior quarter of the pelvis; but, although the left ovary was contained amongst these, they were so densely and deeply adherent that it was deemed unwise to attempt their removal. Below the adherent left tube, in Douglas's pouch, there was a small collection of purulent fluid, containing masses of coagulated lymph. There was a good deal of oozing from separated adhesions, but no wounded vessel was of sufficient importance to need ligature. A large piece of inflamed omentum that had been much soiled by the fœtid pus was ligatured and cut off. A drainage-tube was passed as deeply as possible, and the abdominal wound closed in the ordinary way. The operation lasted two hours.

Description of parts removed.—Both tubes were enlarged, their coats thickened and succulent, a quantity of thick purulent mucus in their canal. The mucous membrane was swollen and discoloured, but the rugæ were very distinct, and there was no ulceration. The fimbriated ends of the tubes were constricted as by a ligature, but were not occluded. The portion of the right tube removed measured 4½ inches in length; it was much contorted, and was larger both in breadth and length than the left tube. The portion of left tube removed was 3 inches in length.

The left ovary was absent.

The right ovary, an inch and a half in diameter was dense from chronic inflammation. At one end of it was a thick-walled cyst, the size of a Tangerine orange, from which the contents had escaped.

The temperature, which immediately before the operation had averaged 99° in the morning, and 100° to 101° in the evening, fell after the operation nearly to normal, the highest record during the first week being 99·4°. The patient required morphia the first two nights. The drainage-tube was removed in forty-four hours. The patient had an action of the bowels (after an enema) and passed urine naturally on the fourth day. The stitches were removed on the ninth day.

On the twelfth day, the patient having complained of a good deal of pain for a day or two, there was observed some abdominal distension, with tenderness and fluctuation near the lower angle of the wound. The lower part of the wound was accordingly reopened for a short distance (½ inch) and exit given to a large quantity of thick, dirty, ill-smelling pus. An india-rubber drainage-tube was inserted, and left in until November 29th, when the discharge had ceased. After that there was no further discharge, except once, viz. on December 4th, when, in consequence of some pain about the lower end of the wound, a probe was passed and a little pus welled out. The temperature between November 4th and December 9th ranged between 97° and 99·2°.

The patient left the hospital on December 10th, looking stout and well.

On February 27th, 1891, she had gained flesh, and was feeling well. She complained of a little pain on the right side of the pelvis, where there was some thickening to be felt in the situation of the pedicle. There was no swelling on the left side or posteriorly.

On June 24th she wrote,—"Since seeing you I have greatly improved in health, and can walk better than I have done for years."

Shortly after this, she consulted me on account of a tumour in the breast. The tumour was removed by one of my surgical colleagues and proved to be a carcinoma. She recovered from the operation, but I heard that she died in November, 1891, from cancer of the stomach.

There is no evidence to show what was the origin of the purulent salpingitis in this case. I am not in possession of the private history of the patient and cannot say whether the mischief was gonorrhœal or septic. Notwithstanding the eminent respectability of the patient, I strongly suspect it was gonorrhœal. Finding a mucous polypus to account for the hæmorrhage, and regarding the pain the patient complained of as mostly, if not wholly neurotic, I very nearly let her leave the hospital without having treated, or even discovered, the active disease going on in the uterine appendages. Even when I found an inflammatory mass behind and to the left of the uterus, I thought it was merely the remains of an old pelvic peritonitis, and might safely be disregarded. It was only when unmistakable symptoms of septic absorption showed themselves that I realised the serious nature of the case. The result fully justified the exploratory operation. There was pus in an ovarian cyst, in both tubes, and amongst the adhesions in Douglas's pouch. The effect of removing all this was highly satisfactory. Unfortunately the patient succumbed to cancerous disease, first of the breast and afterwards of the stomach, before she had enjoyed her renewed health for more than a few months.

CASE 38. *Puerperal peritonitis thirty years ago; no subsequent pregnancy; great pain and discomfort in pelvis since, especially at menstrual periods; symptoms worse during last few months; soft swelling in front of retroverted and adherent uterus, filling up right side of pelvis ; abdominal section ; several serous cysts of right broad ligament ; uterus and appendages bound down by old adhesions ; cysts removed; uterus set free ; death on twelfth day from septic peritonitis.*—A married woman, aged 51, was admitted November 3rd, 1890, on account of very severe pain in the lower part of the back, increased by movement or stooping, also of great pain before and during defecation.

She had borne but one child a year after her marriage, thirty years ago. She states that she was in labour a week and that she was in bed for six weeks after her confinement, with what the doctor said was inflammation. Since then, there has been constant pelvic pain with dysmenorrhœa and a tendency to slight hæmorrhage on the slightest provocation. Menstruation ceased from August 1889 to January 1890. Then there was a profuse flow which lasted a month and from that time to April there was a continuous slight loss accompanied with incessant pain in the back and lower part of the abdomen.

On admission there could be felt a soft swelling equal in size to a man's closed fist in front and to the right of the uterus. The cervix was directed downwards and forwards ; the sound passed backwards three inches.

Per rectum the posterior surface of the body of the uterus could be traced to the fundus, round which the finger could be hooked ; from the cornua a tense band passed upwards and outwards on each side, presumably the upper border of each broad ligament rendered tense.

Abdominal section was performed on November 10th. Occupying the whole of the right side of the pelvis were a number of thin-walled subperitoneal cysts of the right broad ligament, one of which was the size of a large orange. Some of the cysts contained clear serum, others contained serum stained by altered blood. All the cysts were densely adherent to surrounding parts, except anteriorly. With considerable difficulty they were separated, brought into view, and removed by transfixion of the broad ligament beneath them. The body of the retro-

verted uterus was adherent to the posterior pelvic wall by
a number of firm bands which were torn through by the
fingers. The uterus was then straightened and a Hodge's
pessary introduced into the vagina. The tubes and ovaries
were bound down by old adhesions and prevented the body
of the uterus from being fully anteverted. They were
not disturbed. A good deal of bleeding took place from
the separated adhesions. The peritoneal cavity was
douched with hot boracic solution, a glass drainage-tube
inserted and the abdominal wound closed.

Next day there was slight hiccough, retching, nausea,
pain, thirst, and a good deal of abdominal distension. On
the third day there was continual sickness. Much flatus
and a little fæcal matter passed after enemata, but the
distension continued. Drachm doses of magnesium sulphate were tried, and at long intervals injections of morphia. On the fifth day the distended abdomen was punctured in two places with insignificant result. This treatment was repeated next day with no effect. On the
seventh day a copious enema mixed with glycerine was
given. This was followed by the passage of several liquid
motions and much flatus, the distension remaining unrelieved. After this there was no more sickness, and the
bowels continued to act. It was now thought that the
danger had passed, and the patient's bed was moved into
the general ward, but at 4 a.m. on the twelfth day she
complained of very severe pain, and at 8.20 a.m. she died
in a state of collapse.

The temperature was for the most part normal or subnormal throughout; the highest record until a few hours
before death was 99·4°. The pulse varied from 80
to 120.

Autopsy made thirty hours after death by the late Dr.
Gulliver. General peritonitis. The coils of intestine
were glued together by exudation, and there was a considerable quantity of ill-smelling semi-purulent fluid in
the cavity. The inflammation was most intense in the
pelvis. The uterus was adherent to the back of the

pelvis by some old fibrous bands. The left Fallopian tube was occluded at its fimbriated end and formed a cyst containing about an egg-cup full of clear fluid. Ovary normal. On the right side of the pelvis was the stump of the uterine appendages with its ligature.

Had I known that the swelling on the right side of the pelvis consisted merely of a number of sub-serous cysts, I should not have advised an operation. Looking back upon the case, I think probably the best treatment after opening the abdomen would have been to puncture and evacuate the cysts instead of removing them. The after-treatment was based on the supposition that the symptoms were due to simple intestinal paralysis. The autopsy showed that they were really due to septic peritonitis.

CASE 39. *Attack of pelvic peritonitis in March, 1889; hard smooth swelling in right posterior quarter of pelvis pushing uterus forwards and to the left; bursting of abscess into rectum on nineteenth day; recovery with disappearance of tumour and fixation of uterus; readmission November, 1890, on account of pelvic pain and slight purulent discharge from rectum; reappearance of swelling on right side of pelvis; abdominal section; small thick-walled suppurating cyst of right ovary removed; no intra-peritoneal abscess discovered, but subsequent escape of pus from wound; recovery with complete re-establishment of health.*—An unmarried woman, aged 33, an ironer, was admitted November 4th, 1890, on account of pain in the pelvic region and a purulent discharge from the rectum.

Nineteen months ago, viz. on March 19th, 1889, eight days after a normal menstrual period she was suddenly taken ill whilst at her work, with pain in the back and lower part of the abdomen, shivering, nausea, and a profuse discharge of blood from the vagina. She went home at once and to bed, and lay awake with the pain all night. Next day she attempted to resume her work, but had to leave it and go home. She was afterwards seen by a doctor who told her she had inflammation of the bowels with a displacement of the womb, and advised her to seek admission to a hospital.

She was admitted at St. Thomas's under my care April 1st, 1889. The hæmorrhage had by this time ceased, having lasted four days. She still complained, however, of severe pain in the lower part of the abdomen, and she had retention of urine, requiring the use of the catheter. There was a discharge of mucus from the bowel whenever she moved.

The condition found on vaginal examination was as follows:—Uterus displaced anteriorly and fixed; fundus 1¾ inches above top of symphysis pubis a little to left of median line. On right side extending from uterus to lateral wall of pelvis, a hard, uniform, smooth swelling, inseparable from the uterus and rather tender to the touch. No swelling on left side. Immediately behind the supra-vaginal portion of the cervix, a small, hard, irregular swelling. The upper margin of the swelling on the right side 2¼ inches above Poupart's ligament.

On April 6th a discharge of pus took place from the bowel; this continued for forty-eight hours. The pus was thick, yellow, and without odour. The total quantity passed was estimated at 6 to 8 fl. oz. On April 9th there was a discharge from the bowel of clear transparent mucus. On the 18th the patient felt quite well, she had no pain and there was no discharge. On the 30th the physical signs were as follows:—Uterus absolutely fixed; no swelling behind it, but the parts in Douglas's pouch so adherent that the vaginal roof cannot be pushed up. No depression of either lateral fornix, but the whole of the right side of the pelvis occupied by an irregular, fixed, hard mass. Bimanually, no tumour can be felt. Nothing abnormal on left side.

The patient was free from pain; her temperature was normal and had been so since the 8th.

On May 21st the resistance above right fornix and in Douglas's pouch was less marked, though still quite evident. Fixation of uterus less absolute. Patient left the hospital on June 8th.

For the next two months after this she remained quite free from pain or inconvenience of any kind. But about the end of that time she noticed that she had to go to stool more frequently than usual, and she often passed nothing but a small quantity of yellow matter. This continued up to three weeks before her readmission, when the desire to defecate became much more frequent, the matter passed being generally purulent. For the last week she had suffered a good deal of pain whilst at work.

Patient is a thin, sallow, dark-complexioned woman with a badly formed chest. On readmission (November 4th, 1890) there was a hard, smooth swelling felt to right of and behind the uterus, and the evacuations contained pus. The pulse was 72, the temperature normal.

Abdominal section was performed November 12th, 1890. A small, inflamed, tense, and thick-walled cyst of the right ovary containing $3\frac{1}{2}$ fl. oz. of dark, thick, fœtid pus, was with much difficulty separated from the very dense and vascular adhesions which surrounded it on all sides. The cyst was brought into view, punctured with a trocar, partially emptied and removed, together with the inflamed right tube which was closely incorporated in the cyst-wall, but was pervious throughout and did not communicate with the interior. The cyst was single; it measured $2\frac{1}{2}$ inches × $1\frac{1}{2}$ inches; its wall was of the uniform thickness of $\frac{1}{4}$ inch; its cavity was lined by inflammatory lymph. No intra-peritoneal abscess was found or any communication with the rectum discovered; the left ovary was healthy but surrounded with adhesions, which were separated without removing either tube or ovary. The peritoneal cavity was douched and a glass drainage-tube inserted before the abdomen was closed. The operation lasted an hour and a half.

The patient was sick from time to time up to 2 p.m. on the 14th. As there was pus in the discharge, the glass drainage-tube was replaced that day by an india-rubber one. Flatus passed naturally on the 15th. The discharge was slight, the microscope showed it to contain pus. On the 17th the bowels were opened four times after a dose of castor oil; no pus was visible in the evacuations. The stitches were with one exception removed on the 20th; a little ill-smelling pus was then coming from the wound. On December 8th there being little or no discharge, the drainage-tube was finally removed. On December 13th patient was very comfortable; there had been no pain or rise of temperature since the removal of the tube. She sat up in bed on the 10th, got up for the first time on the 15th, and left the hospital well on the 31st. There was no swelling in the pelvis, the uterus was fixed. The highest temperature after the operation was 99·4°, except once (on November 21st) when, after an enema, it reached 100°. From and after December 1st it was uniformly normal or subnormal.

April 18th, 1891.—Patient presented herself on account of a pharyngeal catarrh. In other respects she was quite well. She had gained flesh and her skin had assumed a healthy colour. She had had no pain in the pelvis or discharge from the bowel since leaving the hospital. She had menstruated regularly and been able to do her work easily.

October 17th, 1891.—Applied for help towards the purchase of a belt, the abdominal wall being weak. She has no pain, but when tired has cramp-like sensations in the lower part of the abdomen. She has not menstruated for three months. She is working hard as an ironer two or three days every week from 8 in the morning to 9 at night.

Although no intra-peritoneal abscess was discovered during the operation, the subsequent discharge of pus through the abdominal wound makes it probable that such an abscess existed, the remains of the large abscess which had burst into the rectum eighteen months previously. The inflamed condition of the right Fallopian tube renders it more than likely that the abscess had its origin in suppurative salpingitis, the pus escaping from the tube into the peritoneum. The incomplete evacuation of the abscess when it burst into the bowel would account for the subsequent symptoms and for the infection of the neighbouring ovary. The result of the operation was all that could be desired.

CASE 40. *Pain in left iliac region and temporary rise of temperature on the ninth day after delivery, without discoverable lesion; recurrence of the pain at intervals; pain worse on returning to work; six months after confinement development of a fixed swelling in left posterior quarter of pelvis, with purulent discharge from uterus; diagnosis of salpingitis with pelvic peritonitis; abdominal section; left pyosalpinx with adhesion of tube and ovary; left appendages removed; recovery interrupted by acute bronchitis but otherwise satisfactory; persistence of pelvic pain; no lesion discoverable.*—An unmarried girl, aged 22, a servant, was delivered of a full-term child at the General Lying-in Hospital in April, 1890. The labour was tedious, and delivery was effected by forceps. The perineum was slightly torn. On the ninth day the temperature, which up to that time had not exceeded 100°, rose to 102·6°, and the patient complained of pain in the left iliac region. Dr. Herman made a vaginal examination and found nothing abnormal. Next day the pain had disappeared and the temperature was normal. On the 14th day the patient was sent to a convalescent home, where she remained four weeks. During her stay there she had a recurrence of the pain which was quickly relieved by the application of a blister. After leaving the home, she had a good deal of pain in the back and in the left iliac region, with frequent discharge of blood from the vagina. She was able to do her work, however, until the beginning of September, when the pain became severe, and the discharge continuous and profuse.

She was admitted to St. Thomas's Hospital October 25th, 1890. A purulent discharge was seen issuing from the *os uteri*, both lips of which were the seat of a catarrhal erosion. The cervix was dilated and the interior of the uterus curetted, with the result of bringing away some clots and some fragments of membrane. The pain and yellow discharge continued, and on November 14th an examination was made under anæsthesia. On the left side, anterior to and below the retroverted body of the uterus, was felt a well-defined oblong mass, depressing the left vaginal fornix. The mass was divided by a sulcus into two portions, one a smooth, rounded body, suggestive of an ovary, the other a hard, elongated swelling passing some distance outwards, and situated behind and below the smaller swelling. Nothing abnormal was felt on the right side.

The diagnosis was diseased left Fallopian tube, with pelvic peritonitis and adhesion of tube and ovary to each other and to surrounding parts.

Abdominal section was proposed and agreed to.

The operation was performed on November 19th. The uterus was slightly thicker and larger than normal, somewhat retroverted and inclined to the right side, and connected with the posterior wall of the pelvis by numerous bands of adhesion, recent and easily separated. The left tube and ovary were adherent to each other, to the back of the broad ligament and to other surrounding parts. The tube was thickened by inflammation, and unequally dilated, owing to a sharp bend. The main dilatations were two in number, and were felt to contain fluid. The distal end was occluded. The ovary was normal in size and appearance, but completely enveloped by adhesions. The ovary and tube were removed together. The right tube was normal. The right ovary was normal in size and appearance, but was prolapsed and adherent over its entire surface. These adhesions having been easily separated, the right tube and ovary were left without further interference. There was a considerable amount of oozing from the separated adhesions. The pelvis was well sponged and a glass drainage-tube inserted before closing the abdominal wound. The operation lasted an hour and a quarter.

On opening the diseased left tube it was found to contain thin purulent fluid. The mucous membrane was pale and swollen, but showed no sign of ulceration, new or old. The muscular wall was thickened; it measured a ¼ inch. The ovary was normal.

Convalescence was retarded by an attack of bronchitis, but otherwise she made a good recovery, and was able to be sent to a convalescent home on December 10th.

On January 13th, 1891, she was readmitted and she then made the following statement. Two or three days after leaving the hospital she began to reject her food from half an hour to an hour after each meal. Pain in the back, which was present to a slight extent when she was discharged, became worse, shooting into the right side. She had had attacks of shivering followed by perspirations. Three days after leaving the hospital, the yellow vaginal discharge had reappeared.

Her temperature on readmission was normal. She was examined on the 16th of January, and again under ether on the 28th, with an entirely negative result. No swelling could be found in either posterior quarter of the pelvis. She was accordingly discharged.

A few months later she applied at the Westminster Hospital complaining of pelvic pain. She was examined by Dr. Potter and nothing abnormal was found.

On November 17th, 1891, she presented herself again at St. Thomas's, still complaining of pelvic pain and some metrostaxis. I examined her carefully but could not detect any swelling. The uterus was movable and the posterior quarters of the pelvis free.

This is one of the very few cases in which pain has persisted after removal of diseased appendages. It may, of course, be due to mischief in the remaining and apparently healthy tube, but in the absence of any evidence of peritonitis or alteration in the size and position of the right appendages, I am much more disposed to think that the pain has no pathological significance. I hope I am not doing the girl an injustice if I suggest that her persistent complaints are due to her having discovered that hospital life is easier than the work of a domestic servant.

CASE 41. *Acute pelvic peritonitis seven weeks after fourth confinement; a week later large mass on left side of pelvis and smaller one on right, depressing vaginal roof laterally; after another week, swellings smaller and better defined, thickened and adherent tubes being traceable; development of cellulitis around cervix; gradual recovery; return to household duties for nine months, though in more or less constant pain; recurrence of acute peritonitis; soft mass in right posterior quarter of pelvis with thickened Fallopian tube; diagnosis of diseased right ovary with inflamed tubes and peritonitis; abdominal section: right ovary enlarged and honeycombed with abscesses; right tube occluded and inflamed; pelvic contents matted together; right appendages removed; left fairly healthy, not removed; accidental wound of intestine sutured; escape of fœtid pus from lower angle of wound on tenth day; fæcal stain on one occasion only; recovery with re-establishment of health; small sinus with slight discharge two months after.*—A married woman, aged 25, was first admitted to St. Thomas's Hospital December 2nd, 1889. She had been confined of her fourth child seven weeks previously. After the confinement she suffered severely from after-pains and headache, and was kept in bed for fourteen days. The headache persisted, and although she went about the house she did not go out-of-doors. Some hours before her admission, she was seized with sudden abdominal pain and faintness. This attack she attributed to having got her feet wet two days previously, whilst washing clothes in the yard.

She was a stout, pale, anæmic, despondent woman of feeble intelligence. Her urine contained one sixth to one tenth albumen.

The fundus uteri was 4 inches above the pubes and 2 inches below the umbilicus. There was a sense of resistance in left iliac region, bounded above by a well-defined margin on a level with the anterior superior iliac spine. There was tenderness in the right iliac fossa, without definite swelling or sense of resistance. The percussion-note was absolutely dull from the top of the pubes to a line 3 inches above that.

On vaginal examination the uterus was found fixed, the cervix shortened, the os patulous. The left fornix was depressed by a firm slightly elastic mass, continuous with the mass felt in the abdomen. The right fornix was slightly depressed, by a similar though less easily definable mass. There was no fulness or depression of the pouch of Douglas, but high up a firm band could be felt stretching across behind the upper part of the cervix. This band was still more distinct on examination *per rectum*.

The temperature, which on admission was 101°, varied on the 3rd December between 100·6° and 102·6°.

,,	4th	,,	,,	. 100·4°	,,	103·2°
,,	5th	,,	,,	99°	,,	101·2°
,,	6th	,,	,,	100°	,,	101·8°
,,	7th	,,	,,	99·6°	,,	102·4°
,,	8th	,,	,,	98·4°	,,	102·8°
,,	9th	,,	,,	97°	,,	103·4°
,,	10th	,,	,,	98°	,,	100°

After which it was usually normal, and did not exceed 100·6° up to the time of the patient's discharge on the 26th January.

On December 17th (a fortnight after admission) the firm mass in the left iliac fossa had disappeared. The left vaginal fornix was slightly depressed by a firm mass which, bimanually, could be differentiated into, anteriorly, a thickened and contorted Fallopian tube, traceable from the cornu of the uterus outwards and curving round to the back of the broad ligament, and posteriorly, a larger softer mass, thought to be the ovary. High up, behind the cervix, could be felt a fixed, firm, swelling continuous with the adherent mass already described as occupying the left posterior quarter of the pelvis. The right appendages were not felt through the vagina, but, *per rectum*, the right tube, thickened, could be felt bent upon itself and turning down behind the uterus.

The urine still contained one tenth albumen. On January 10th, 1890, the uterus was in normal position, its mobility impaired. Masses felt on both sides of the pelvis, apparently consisting of broad ligament and appendages intimately matted together. The mass on the right, higher up than that on the left, extended outwards and backwards to the pelvic wall. On the left the tube ran out and back and then curved downwards behind the uterus, closely adherent to the mass round which it curved. The lateral fornices were depressed; the supra-vaginal portion of the cervix was completely surrounded by a hard collar. *Per rectum* a depression could be felt in the middle line above the cervix, and, higher up, a firm transverse band, causing a projection in the rectum. From this band tense bands could be felt, diverging like the arms of the letter V, and passing upwards and backwards. The right arm of the V was more distinct than the left, which was interrupted by a rounded prominence.

On January 22nd, 1890, the swellings in the pelvis were all found smaller, and on the 26th the patient went home.

The patient was readmitted on November 12th, 1890. She then stated that she had been able to do her housework and look after her children ever since she left the hospital, though she had never felt really well, and had suffered from time to time from pain in the pelvis and thighs, especially on the right side. On November 8th she was taken ill with vomiting and very severe pain in the abdomen and right groin shooting down the thigh. Since then she has perspired profusely at night.

On vaginal examination there was found an ill-defined soft mass behind and to the right of the uterus; and a smaller, harder, and more irregular mass on the left. The uterus was normal in position. There was a trace of albumen in the urine.

The diagnosis was a diseased and enlarged right ovary with inflamed and adherent Fallopian tubes.

On November 28th, 1890, abdominal section was performed. The pelvic viscera were found matted together, omentum and large intestine being also involved. A band of omentum was adherent to the bladder, and a broader one to the parts behind the uterus. These were tied and divided. To the right of and behind the uterus was a soft rounded mass, which, after careful separation of adhesions, was brought into view and seen to be the right ovary diseased and enlarged. With the ovary was removed the inflamed and thickened right tube. The left appendages being fairly healthy, it was decided not to remove them. A thickened coil of large intestine dipped down to the floor of the retro-uterine space to which it was intimately adherent. This having been separated and brought into view, it was found that, during the process of separation, the coats of the intestine had been torn, leaving an aperture large enough to admit the tip of the finger, through which the mucous membrane protruded. This rent was closed by four Lembert's sutures of fine silk. The tip of the appendix vermiformis was also adherent to the floor of the retro-uterine space; this was left undisturbed. The pelvis was now cleansed with sponges, a glass drainage-tube inserted, and the abdominal wound closed.

Description of parts removed.—The right ovary measured $2\frac{1}{2}$ inches by $1\frac{3}{4}$ inches. On section it was found to be honeycombed with spaces, containing thick yellow pus. The portion of right tube removed was $2\frac{1}{2}$ inches in length. Its wall was $\frac{1}{6}$ inch thick; the fimbriated end was occluded. The mucous membrane was swollen and œdematous. There was no ulceration and no pus was found in the tube. The mesosalpinx was thickened. Mr. Shattock reported that the ovary had very much the appearance of tuberculous disease. A portion was put aside for microscopical examination, but appears to have got misplaced, as it could not afterwards be found.

The convalescence was somewhat prolonged. The drainage-tube was removed in twenty-four hours. On the third day some blood escaped from the rectum. On the fourth, the urine contained a good deal of blood. On the fifth day two fluid ounces of blood passed from the rectum. The stitches were removed in a week. On the following day, there was a fæcal stain on the dressing ; nothing of the kind was seen again. On the tenth day a little fœtid pus escaped on probing the lower angle of the wound, and on the twelfth day there was a more free discharge of pus of the same character, but again without any admixture of fæces. After this, the temperature was normal, the discharge was very slight and less offensive, and the patient improved in every way. She was discharged on the 24th January, 1891. She had gained flesh and had a good appetite. The sinus had not quite healed ; the discharge was very slight, and not offensive. On February 17th the sinus was still discharging; the general health was very good. The first menstruation occurred February 13th to 16th.

Whatever the nature of the ovarian abscesses in this case, it seems quite certain that the earlier attack of pelvic inflammation, in which the tubes, the peritoneum, and the pelvic connective tissue were all involved, was of a septic character. My own belief is that the ovary became the seat of suppuration at that time, as part of the septicæmic process, and that the disorganization of the ovary had been going on ever since, without producing very definite symptoms, until, on some slight provocation, a fresh attack of acute peritonitis occurred and the patient became very seriously ill. This is a very common experience. A patient often goes about for months with pelvic suppuration, provided the pus be well shut off from the peritoneum. But she is always on the brink of a precipice, liable at any moment to have her life imperilled from fresh inflammation or from the advance of the destructive process.

CASE 42. *Uterine hæmorrhage followed by occasional pain in the pelvic region in a girl of twenty ; continuance of symptoms for two months ; swelling in abdominal wall, and soft elastic mass in right posterior quarter of pelvis, thought to be a hæmatoma ; no diminution after a month's rest ; abdominal section ; abscess (tubercular) in sheath of right rectus abdominis ; miliary tubercle of entire peritoneum, without peritonitis ; soft mass beneath peritoneum covering posterior part of floor of pelvis on each side ; abscess in abdominal wall evacuated ; abdomen closed ; recovery ; no further symptoms beyond wasting ; twelve months later health completely restored.*—A girl aged 20, engaged as a mother's help, applied for treatment in the out-patient department of St. Thomas's Hospital on account of hæmorrhage which had been going on for two months. There was no obvious cause for the hæmorrhage, menstruation having been previously quite regular. It commenced with a profuse discharge in August, 1890, two weeks after a period, as she was carrying coals in the usual way. She had no pain at the time, but has since occasionally had pain in the lower part of the abdomen.

She was admitted on the 25th October, 1890. She had a healthy appearance ; her colour was good, and she walked as though nothing were amiss. The heart and lungs were normal. There was a small smooth swelling, about the size of a pigeon's egg, apparently in the abdominal wall, just above the right pubic spine.

A vaginal examination was made on October 29th under ether. The uterus was of normal size ; the body directed somewhat towards the left, cervix towards the right. To the right of the uterus and on a plane posterior to it, was a soft ill-defined swelling. This was thought to be a hæmatoma of the broad ligament, and it was decided to watch it.

On November 15th the mass had rather increased than

diminished, and it was decided to make an exploratory incision. There has been no hæmorrhage since October 28th.

On November 22nd abdominal section was performed. The lump above the right pubic spine was cut into in making the incision, and was found to be an abscess in the sheath of the right *rectus abdominis*. About 1½ fl. oz. of thick curdy pus was evacuated. On opening the abdominal cavity, the parietal and visceral layers of peritoneum were seen to be everywhere studded thickly with miliary tubercles. A large, soft, elastic mass was felt deeply down in the posterior part of each side of the pelvis. The structures implicated could not be differentiated. It was considered inadvisable to interfere with these swelling, and a drainage-tube having been inserted, the abscess-cavity in the abdominal wall was thoroughly scraped and the abdomen closed.

The patient recovered from the operation without a bad symptom, and left the hospital on the 20th December. After this, she became very thin and weak.

Exactly twelve months after the operation the patient was examined by Dr. Herbert Hawkins, Assistant Physician to St. Thomas's Hospital. She was looking very well and had completely regained her strength. She presented no physical signs of disease either in the chest or abdomen. Shortly before this I had made a vaginal examination and found little or no swelling; the uterus was in its normal position.

October 22nd, 1892.—Is again losing flesh and feeling weak. No definite signs of disease.

There can be little doubt that the masses in the pelvis were of tubercular origin; their probable seat being beneath the peritoneum lining the floor and posterior wall of the pelvis. I did not open them because I did not see how, at such a depth, they could be efficiently drained, and, in the presence of disseminated tubercle of the general peritoneum, it did not seem justifiable to attempt any radical operation for their removal. For some months the pale and wasted appearance of the patient suggested general tuberculosis, and her present healthy look and improved condition have certainly filled me with surprise. The case is one of much interest and importance in connection with the curability of peritoneal tubercle. Its bearing on this question has been dealt with by my colleague, Dr. Hawkins, in a paper published in the 'St. Thomas's Hospital Reports,' New Series, vol. xx, p. 25.

CASE 43. *Pelvic pain, commencing two months after marriage, gradually increasing for two years; loss of flesh and of strength; entire inability to work for eight months, and for one month entire confinement to bed; pelvis filled with a lobulated swelling pushing uterus forwards and to left, diagnosed as double pyosalpinx; history of gonorrhœa in the husband a few months before marriage; abdominal section: both tubes greatly distended with pus; ulceration of their mucous membrane; tubes removed; shock of operation severe; uninterrupted convalescence; complete restoration to health with regular and normal menstruation.—* A thin, anæmic, highly nervous woman, 25 years of age, was admitted December 9th, 1890, complaining of pain in the lower part of the abdomen, with loss of flesh and appetite dating from two months after her marriage in August, 1888. There had been no pregnancy. Menstruation which, before marriage, had been regular and almost painless, had since been irregular and preceded by considerable pain. The pain in the intermenstrual periods came on gradually, and was worse after standing and after exertion. It was felt not only in the abdomen but in the back and thighs, and was accompanied by increasing weakness and inability to do her work. She first noticed a vaginal discharge about nine months after her marriage; sometimes it was white, but more often yellow and offensive. She consulted a doctor who diagnosed displacement, whereupon she was treated for many months by different kinds of pessaries without benefit. In August, 1890, she consulted Dr. Gervis, who told her that her womb was not displaced, but that she was suffering from inflammation. Dr. Gervis again saw her two days before her admission, and as she was in a less satisfactory condition than on the previous occasion, he advised her to come into St. Thomas's Hospital.

Patient had been unable to do any work for eight months, and for the last four months had been obliged to lie down almost entirely. For the past month she had been in bed.

No abdominal swelling was present; and no tumour or undue sense of resistance.

The uterus was moveable and inclined slightly to the left. There was a lobulated mass situated behind and to the right of it, with a well-marked sulcus between the lobes where they met behind the uterus. The vaginal roof on both sides was somewhat depressed by the pelvic mass.

The temperature was normal during the week following admission, except on December 15th and 16th, when it rose to 100°.

The diagnosis was enlarged and suppurating Fallopian tubes—double pyosalpinx, probably gonorrhœal.

The husband had suffered from gonorrhœa eight months before marriage, and believed himself, at the time of his marriage, to be cured.

Abdominal section having been proposed and agreed to, the operation was performed on December 17th, 1890.

The pelvis was filled by a large mass, consisting of the two Fallopian tubes, greatly enlarged, curved upon themselves, and universally adherent to the parts around, viz. to the uterus, to the broad ligaments, to each other, to omentum, to intestine, and to pelvic wall. The two tubes were separated and removed. The process of separation was prolonged and difficult. The tube in each case underwent a slight tear, permitting the escape of a little thick pus. The rents were quickly clamped. A good deal of oozing took place from the raw surfaces. The ovaries were not seen. Two processes of thickened peritoneum were ligatured and removed, and an enlarged mesenteric gland the size of a pea was also removed for examination. No miliary tubercles were seen, but there were one or two suspicious subperitoneal thickenings on the intestine. One of these was suppurating and burst. The peritoneal cavity was abundantly douched with hot boracic solution and then sponged, and, after a glass drainage-tube had been inserted, the abdominal wound was closed. The operation lasted two hours.

Description of parts removed.—The right tube had a circumference of 4½ inches, its length was 4¼ inches; its breadth 1½ inches; its width when laid open 3 inches. The left tube had a circumference of 6½ inches; its length was 6 inches; its breadth 3 inches; its width when laid open 4 inches.

The surfaces were red and vascular and covered in places with shreds of adhesion. The contents of both consisted of very thick pus with some mucus. The mucous membrane was ulcerated throughout.

The mesenteric gland, on section, proved to contain either cheesy tubercle or inspissated pus. It was examined microscopically by Mr. Shattock, who reported that there was no trace of tubercular disease.

The collapse after the operation was very severe and prolonged, but after reaction had set in convalescence progressed without interruption. No suppuration occurred from the wound, and the temperature on no occasion exceeded 100·2°.

On January 9th, 1891, a vaginal examination showed the uterus central in position, the fundus adherent to the anterior abdominal wall. There was no swelling behind or to the right of the uterus; the base of the left broad ligament was thickened, slightly depressing the vaginal roof. For three or four days before the patient went home, there was a purulent vaginal discharge.

On March 6th the patient attended, looking and feeling well; she had gained flesh and had no pelvic pain or discomfort. Both posterior quarters of the pelvis were

CERTAIN CASES OF PELVIC PERITONITIS. 121

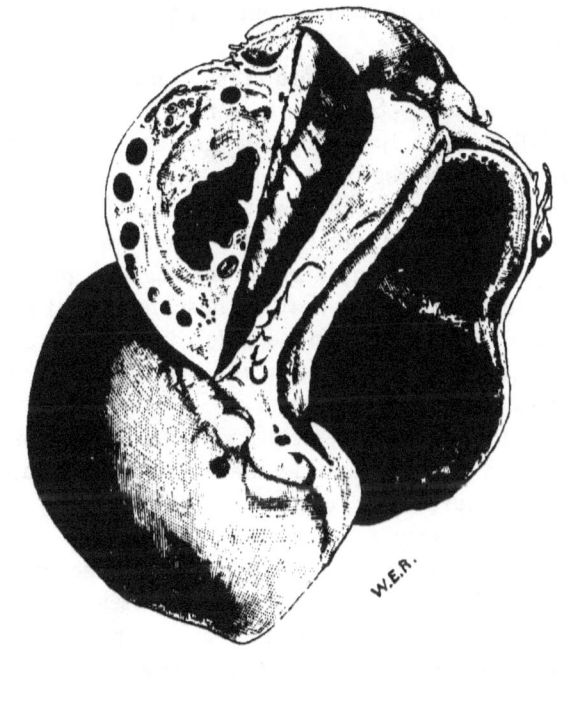

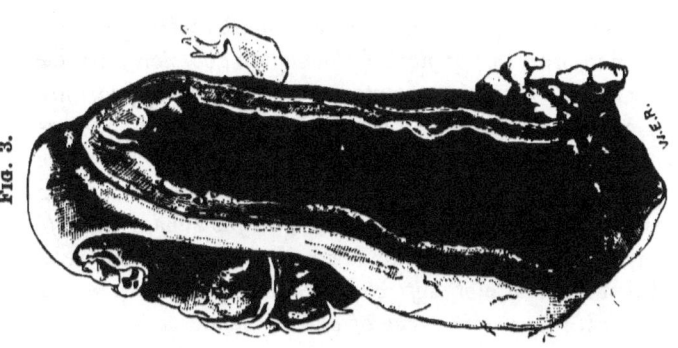

Fig. 3. Pyosalpinx due to gonorrhœa. Both tubes enormously enlarged, and filled with pus. The walls greatly thickened by chronic inflammation.

Fig. 4. The irregular dilatations of the tubes are well seen. (Case 43.) (The engraving is reduced from a pencil drawing by Mr. W. E. Roth, taken after the specimen had been mounted for the St. Thomas's Hospital Museum. Three-fourths of the natural size.)

free. The vaginal mucous membrane was bright red and was covered with purulent discharge. The discharge issuing from the os uteri was clear and transparent. She had menstruated twice.

I last saw her January 5th, 1892, more than a year after the operation. She was free from pain, quite well, and in full work. She was still gaining flesh. Menstruation was quite regular.

This case was one of the most satisfactory in the series. The tubes were the largest I have ever yet met with; they simply consisted of large, tense bags of pus. All the evidence goes to show that the mischief was of gonorrhœal origin. I am glad to be able to report as to the patient's condition a year after the operation. The restoration to health is complete.

The rapidity of convalescence after so prolonged an operation, and notwithstanding the unavoidable escape of some of the purulent contents of the tubes into the pelvis, was singularly satisfactory.

It will be noted that on four out of the seven days that elapsed between the patient's admission and the operation the temperature was normal or subnormal. The bearing of this fact on the diagnosis of pelvic suppuration is obvious.

CASE 44. *Pelvic pain for eighteen months with progressive weakness and loss of flesh following an abortion ; tender swellings in both posterior quarters of pelvis ; disappearance of signs after a few weeks of hospital treatment ; recurrence of pain immediately after discharge ; readmission three months later after missing three menstrual periods; large sausage-shaped swelling in situation of right tube, with soft mass in Douglas's pouch ; abdominal section : right tube distended with firm clot, soft clot protruding from open end of tube, continuous with mass of clot in pelvic cavity ; left tube occluded; appendages removed ; recovery interrupted by a pelvic abscess.*—A married woman, aged 25, was admitted into St. Thomas's Hospital on August 11th,

1890, on account of sickness, loss of flesh, and pain in the lower part of the abdomen, especially after standing or walking. Her symptoms dated from an abortion at the fifth month, a year and a half previously. She had been married seven years. A year after marriage she had a still-born child at seven months, and fourteen months later she had another child born at seven months. After this she had two abortions, each at five months. She remained in bed a fortnight after the latter of these abortions, and had to go back again to bed almost immediately, on account of the symptoms above enumerated. She had also suffered, ever since that time, from a thick yellow vaginal discharge, and from pain on micturition.

The patient's husband, a sailor, was in the surgical wards with a severe stricture of the urethra at the same time that she herself was in Adelaide Ward.

The abdomen was resonant throughout; no tumour was perceptible.

On August 19th, the patient was examined *per vaginam* under ether. The uterus was normal in size and position and was fairly moveable. A firm, elastic swelling was felt on the left side in the situation of the left Fallopian tube; the diameter of the swelling was estimated at ¾ inch. There was also a swelling on the right side of the pelvis, less clearly defined, thought to be the tube bent upon itself. In the posterior cul-de-sac there was a small, hard, moveable body to the left, and an ill-defined, irregular swelling high up to the right, probably the distal end of the right tube. The patient, who on admission looked worn and ill, had now greatly improved in appearance and suffered less pain.

On August 27th, the physical signs in the pelvis had altered remarkably for the better. The uterine appendages could be made out distinctly on each side, nearly of normal dimensions, those on the left being perhaps a little larger than on the right. There was no tenderness on either side. High up in Douglas's pouch there was a tender spot, but bimanually, with a finger in the rectum, the fingers can be made nearly to meet behind the uterus.

The patient was discharged on August 30th.

On December 15th, she was readmitted, having been laid up ever since leaving the hospital. She had not menstruated since the first week in September.

There was a mass in the hypogastric region rising two inches above the symphysis, and having a breadth of two inches. There was no corresponding prominence of the abdomen. The cervix uteri was depressed, the body displaced forwards and anteflexed. Behind and to the right of the uterus was a swelling, even, soft and tender, extending outwards from the right cornu of the uterus and terminating posteriorly behind the supra-vaginal portion of the cervix on the left side. In the position of the left broad ligament, a thickened tube could be felt along its border. The mass in the retro-uterine pouch caused a depression of the posterior part of the vaginal roof.

Abdominal section was performed on December 18th. The right tube formed a sausage-shaped mass and was adherent to the surrounding parts. From its fimbriated end, which was open, protruded a large quantity of dark firm clot. The left tube was occluded at its distal end, but was otherwise normal. Both tubes were removed with the adjacent ovaries. The clots lying in the pelvis were cleared away, and the cavity was well douched.

Description of the parts removed.—The portion of the right tube removed was 4 inches long, and $5\frac{3}{4}$ inches in circumference. It was filled with old, firm, partly decolourised clot, closely adherent to its walls. From its open mouth a quantity of firm black clot projected. Enlarged veins, filled with clot, were seen beneath the mucous membrane.

The portion of the left tube removed was 2 inches in length and $2\frac{1}{2}$ inches in circumference. Its distal end was occluded. Otherwise it was normal, except for a subperitoneal cyst.

A week after the operation, the patient began to complain of pain in the pelvis and the temperature rose to 101°. On the eleventh day, there was some abdominal distension and a swelling could be felt *per vaginam*, to the left of the uterus. Two days later, the lower end of the wound was bulging. On passing a probe and exercising a little force, an abscess was reached, and about 4 fl. oz. of dark-brown highly offensive fluid, of putrid odour was evacuated. Next day the temperature had come down from a maximum of 102·4° to one of 98·2° and the pain had disappeared.

On the 18th January, 1891, the patient left the hospital with a normal temperature and a very slight discharge.

The sinus finally closed on February 7th.

On March 21st, three months after the operation, the patient attended, complaining of flushings of the face and of some pelvic pain during the last few days. There was a hard, tender spot at the site of the pedicle on the right side, and a small nodule of hardness on the left side of Douglas's pouch. The uterus was freely moveable.

The probable explanation of this case is that the attack of salpingitis and pelvic peritonitis from which the patient suffered in August left her with damaged appendages; that she became pregnant soon after leaving the hospital; that the gestation was tubal; and that it ended in tubal abortion. As no remains of an ovum were found, this view is, of course, hypothetical. Had it been placed beyond doubt that hæmatosalpinx was the result of conception, I should have classed the case under the head of extra-uterine gestation, and not included it in the present series.

CASE 45. *Pain in the left iliac region with irregular and painful menstruation, and purulent intermenstrual discharge for two years; fixed swelling in left posterior quarter of pelvis size of small orange; less defined and more flaccid swelling on right side; abdominal section; pyosalpinx on left side; hydrosalpinx on right; removal of both tubes and both ovaries, the latter being normal but intimately involved*

9

in the adhesions; rapid recovery from the operation; persistence of pelvic pain; development of tense cyst in left broad ligament; removal by enucleation at King's College Hospital; pain still unrelieved.—A thin, delicate-looking, anæmic married woman, aged 32, was admitted into St. Thomas's Hospital, December 15th, 1890.

Her marriage took place in 1877. She has had one child, stillborn, a year after her marriage. The labour was normal, and she was able to get up in a fortnight, but she has never felt strong since.

Her present illness commenced two years ago with a yellow vaginal discharge, bearing-down, painful micturition, and pain in the left iliac region, felt most after standing and walking. From that time she has been continuously under medical treatment, using vaginal injections. There had been during the whole of the past two years irregular and painful menstruation. A week ago, she began to suffer from diarrhœa and a very severe shooting pain in the lower part of the abdomen, shooting down the left thigh. She was so weak and ill that she was attended by a doctor at her own home; and being no better after a few days came up to the hospital.

On admission, the treatment was directed to the dysmenorrhœa, which at that time was what she chiefly complained of. The cervical canal was dilated with graduated metallic bougies. This occasioned a good deal of pain, and the patient became faint and covered with perspiration. She complained of much pain in the left iliac region during the next few days, and on December 26th the resident in charge made a vaginal examination. He noted that the uterus was moveable and slightly retroverted; behind and to the left side of the uterus was a rounded elastic swelling equal in size to a small apple, slightly depressing the vaginal roof on the left side. Nothing abnormal was detected on the right side.

The temperature was usually normal; one day it was 99° and another 99·4°; these were the highest records since her admission.

Fig. 5.

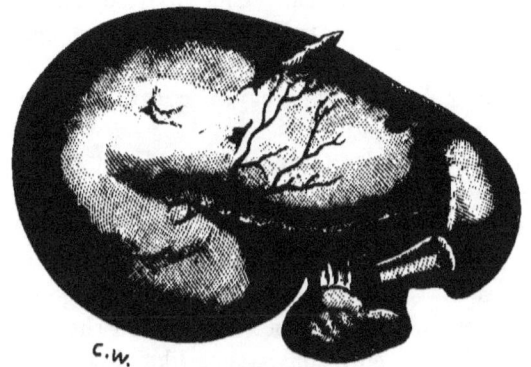

Pyosalpinx (left).

Fig. 6.

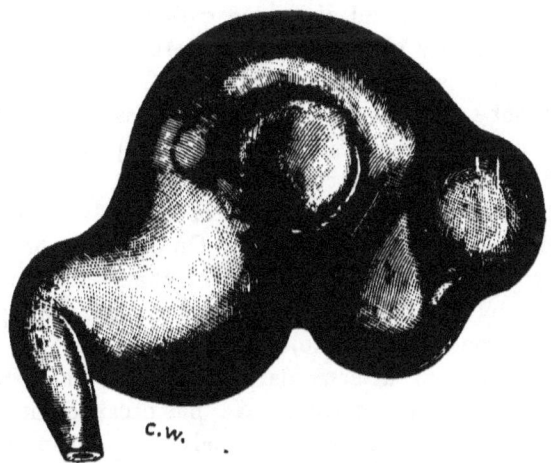

Hydrosalpinx (right).

Pyosalpinx of left side and Hydrosalpinx of right in the same patient. The hydrosalpinx is secondary, being the result of occlusion of the distal end of the tube by peritonitis set up by the pyosalpinx on the opposite side. Natural size. (Case 45.)

On discovering the morbid condition of the left uterine appendages, the resident advised the patient to remain in the hospital until my return. She insisted, however, on going out and left the hospital the same day (December 26th).

On reaching home she was in great pain and was obliged to keep her bed.

She was readmitted on January 1st, 1891.

On January 3rd she was examined under an anæsthetic. The uterus was retroverted, directed to the right, and freely moveable. To the left and posteriorly was felt a distinct mass of the size of a tangerine orange; it was separated from the uterus by a sulcus. On the right side was a thickening of soft consistence like that of a coil of intestine, in the situation of the broad ligament.

Abdominal section was performed on January 8th. On the right side, covered by adherent omentum, was found a tense dark-coloured swelling, with thin walls, consisting of the right tube, doubled and coiled upon itself, occluded at its fimbriated end, and distended with clear fluid. The swelling lay partly in front of the uterus. Behind and adherent to it was the normal ovary. After separating the right tube and ovary from their adhesions, and removing them, the separation of the left appendages was proceded with. They formed an adherent mass which occupied the left posterior quarter of the pelvis and extended into the retro-uterine pouch. On bringing the mass into view it was found to consist of the normal ovary surrounded by the much elongated and twisted Fallopian tube which was of an opaque, yellowish-white colour, and distended with fluid, afterwards proved to be pus. Both tube and ovary were removed. No pus escaped into the pelvis during the operation. The pelvis was cleansed by sponging, a glass drainage-tube inserted and the abdominal wound closed.

The removed portion of the right tube was, when uncoiled, $4\frac{1}{4}$ inches long and an inch in diameter. The mass, before being uncoiled, measured 2⅛ inches in length and 2 inches in breadth.

The removed portion of the left tube was, when uncoiled, 5½ inches long, and ¾ inch in diameter. The mass before being uncoiled measured 2¾ inches × 1¾ inches.

The fluid in the right tube was thin serum; that in the left was thick, yellow pus. The walls of the former were attenuated; those of the latter much thickened.

The patient made a rapid and uninterrupted recovery and left the hospital well on January 31st.

Her subsequent history is somewhat interesting.

On February 24th, she had improved considerably in health, but complained of some pain in the left iliac region. On vaginal examination a swelling equal in size to and closely simulating a tense and full-sized ovary was felt lying against the left lateral wall of the pelvis. Nothing abnormal was detected on the right side.

Had I not known that both ovaries had been removed, I should have regarded this little swelling as the left ovary, rendered tense by a small cyst. Anyway, I regarded it as of little or no importance, and did not propose to take any steps for its removal.

The patient, after a little time, applied at King's College Hospital still complaining of pain in the left side. She was admitted under the care of Dr. Hayes, who very courteously communicated with me. I gave him the history of the patient so far as I knew it. On July 15th, 1891, I received a letter from him informing me that he had that morning operated upon her, and removed a cyst, the size of a large hen's egg, from the left broad ligament. It was enucleated and removed without rupture.

On November 13th, Dr. Horrocks wrote to tell me that the patient had come under his care at Guy's Hospital, and to ask me if it was correct that I had removed one of her ovaries some months ago, and if so, what was the condition of the one left behind. I gave him the particulars of the previous operations. He has since informed me that my report prevented him from reopening the abdomen in search for an ovary that had already been removed. He tells me that the patient declares that she has just as much pain as she had before any operation was performed.

Had I removed the appendages in this case merely on account of pain, the after-history just recorded would have obliged me to confess that the operation had failed in its object. Fortunately for my peace of mind, it was not so, and all that the after-history really shows is that a neurotic condition co-existed with a definite serious lesion, and that the removal of the part actually diseased has not cured the neurosis.

With reference to the fact of there being a pyosalpinx on one side and a hydrosalpinx on the other, I believe the explanation to be that the latter was a mere incident in the course of the pelvic peritonitis set up by the pyosalpinx, being as it were a retention-cyst due to the occlusion, by peritoneal adhesions, of the fimbriated end of the tube.

CASE 46. *Purulent vaginal discharge for four years; acute pelvic peritonitis after a debauch and exposure to wet; tender swelling in left side of pelvis displacing uterus to right; thickened tube in front of swelling; abdominal section: interstitial salpingitis on left side; blood-cyst of left ovary; left tube and ovary removed; right appendages normal; recovery.*—An unmarried girl, aged 23, a machinist, was admitted into St. Thomas's Hospital, January 3rd, 1891, on account of abdominal pain of three weeks' duration, and a yellow vaginal discharge that she had had for four years.

She stated that on the 12th December, 1890, and again on the following day, she had got her feet wet, and that in the afternoon of the second day she was attacked with "crampy" pains in the lower part of the abdomen. Two days later she took to her bed, and had remained there up to the time of her admission. A few days before this attack she went out for the evening with a discarded suitor, and had something to drink. On awaking next morning she found herself very sore, and noticed some blood on her linen. She remembered that her companion had taken liberties with her, but was not aware that actual intercourse had taken place.

She was a pale, poorly-nourished girl, deeply marked by smallpox. Her skin was hot and dry; her temperature at 8 p.m. on the day of admission was 103·6° and at midnight 104·2°. She had no rash. The tongue was thickly coated with white fur. The abdomen was rigid but not distended; there was no tumour perceptible.

Next day she was much better. The temperature was 102·4° at 4 a.m.; 101° at 8 a.m.; 100·6° at noon; 99° at 4 p.m. and 101° at 8 p.m. After that, the temperature became gradually lower, and on January 10th it was normal.

A vaginal examination was made on January 9th, having been deferred on account of menstruation. The uterus lay a little to the right. A tender swelling the size of a small apple could be felt on the left side of the pelvis, causing some depression of the vaginal roof. In front of the swelling, immediately beneath the abdominal wall, was a tense band running horizontally outwards, thought to be the thickened Fallopian tube. Nothing abnormal was detected on the right side.

Abdominal section was performed on January 15th. The left tube was found thickened and adherent, embracing the ovary, enlarged to the size of a pigeon's egg. During the process of separation a cyst in the ovary was accidentally ruptured, giving exit to a small quantity of dark fluid blood. The left broad ligament was somewhat thickened by cellulitis. The left tube and ovary were removed. The right appendages were healthy. The pelvic cavity was sponged and the abdomen closed without drainage.

The portion of left tube removed, when uncoiled, measured 3¼ inches in length and ¾ inch in its greatest diameter. It walls were three times the normal thickness. The mucous membrane was healthy. The fimbriated end of the tube was open and there was no fluid of any kind in the canal.

The patient made an uninterrupted recovery and was discharged well on February 7th, both sides of the pelvis being free from abdominal swelling.

This was an example of interstitial salpingitis, probably of old date and due in the first instance to an endosalpingitis, which had now disappeared. It seems most likely, from the history, that the inflammation was of gonorrhœal origin, the acute attack of pelvic peritonitis, immediately preceding admission, being probably excited by the debauch she described, and aggravated by subsequent exposure to wet.

The main part of the swelling consisted of the cystic ovary, which had been the seat of a more or less recent hæmorrhage.

The strictly unilateral character of the inflammation was somewhat unusual.

CASE 47. *Sudden attack of pain in pelvis two months after confinement five years ago; recurrent attacks of a similar character ever since; continuous pain in left iliac region for a month, obliging patient for the most part to keep her bed; no menorrhagia or vaginal discharge; temperature normal; large mass occupying left posterior quarter of pelvis; indistinct thickening on right side; no depression of vaginal roof; abdominal section: pelvic contents matted by adhesions; outer half of left tube distended, and filled with clot continuous with a small intraperitoneal hæmatocele; hydrosalpinx on right side; ovaries cystic; ovaries and tubes removed; recovery.*—A married woman, aged 31, employed as a charwoman, was admitted into St. Thomas's Hospital January 8th, 1891.

The catamenia had not commenced until the age of seventeen and were habitually scanty. The patient married at twenty-four, and had one child at full term a year afterwards. She recovered well from the confinement, but two months afterwards she was suddenly seized whilst walking with pain in the lower part of the abdomen, especially on the left side. The pain was very severe and extended into the thighs. It soon disappeared, but, ever since, patient has been subject, especially after over-exertion, to attacks of pain of a similar character, accompanied with headache, nausea and faintness. The attacks do not appear to have had any special connection with the catamenia. During the last month they have become more frequent, occurring every two or three days, and patient has also suffered from continuous aching pain in the left iliac region and in the back. She went to bed of her own accord, and then sent for her doctor, under whose care she has been for three weeks. She could not remain altogether in bed, as she had to attend to her sick husband, but she was quite unable to follow her usual avocation. She has been losing flesh for the past six months. There has never been any menorrhagia or vaginal discharge.

Her appearance is that of a woman of healthy constitution; she has a fair complexion; a good colour in her cheeks and a cheerful disposition. Her temperature is normal. On vaginal examination, there is felt a large mass directly continuous with the left cornu of the uterus and filling the left posterior quarter of the pelvis. The mass is hard and nodulated posteriorly and terminates behind the uterus in Douglas's pouch. There is no depression of the left vaginal fornix. There is some ill-defined thickening on the right side of the uterus. The right vaginal fornix is not encroached upon. The uterus is normal in length, anteflexed, and displaced to the right of the median line. On withdrawing the examining fingers, they are seen to be stained with fluid of a brownish-red colour, evidently altered blood.

Abdominal section was performed on January 22nd, 1891. Both tubes were dilated and universally adherent, their distal ends lying firmly matted in the retro-uterine pouch. In separating the left tube, the inner half of which was of normal size, the outer half expanded in a funnel-shaped form, a small intraperitoneal hæmatocele was opened, containing firm, dark clot. Precisely similar clot filled the expanded outer half of the tube, and protruded from its dilated extremity into the hæmatocele, which was hemmed in on all sides by adhesions and was about equal in size to a Tangerine orange.

The right tube and ovary being involved in the mass behind the uterus, were now freed from their adhesions to allow of the more complete separation of the left tube. Both ovaries were enlarged and cystic, being equal in size to a pigeon's egg. The left tube and ovary were now removed. The left tube on being laid open measured ¼ inch across at its narrower portion, and an inch at its dilated extremity.

The right tube was dilated and occluded, forming a hydrosalpinx. After removal it measured while still unopened 2½ inches in length, 1½ inches in its greatest breadth. Its closed end measured 1 inch × 1½ inches.

On a coil of small intestine which was adherent in Douglas's pouch, there was a patch of adherent bloodclot about the size of a sixpence.

The peritoneal cavity was flushed, a drainage-tube inserted and the abdominal wound closed.

The patient made a good recovery. On the thirteenth day a little pus was noticed on the dressing, and on making gentle pressure a quantity of inoffensive pus escaped from the lower angle of the wound. There was a slight discharge for three or four days, and the wound then healed. The patient left the hospital well on the 25th of February. There were some irregular hard lumps to be felt behind and to the left of the cervix, evidently connected with the pedicle on that side. They gave no pain and were not tender.

The order of pathological events in this case is not easy to trace. From the history and physical signs I expected to find a pyosalpinx on the left side. The swelling consisted instead of a hæmatosalpinx communicating with a small hæmatocele. Whether this was an early tubal abortion is matter of conjecture. No evidence of the remains of an ovum was detected. The hydrosalpinx was evidently secondary to the peritonitis, due to the sealing up of the fimbriated end of the right tube by inflammation, and the formation of a quasi-retention cyst.

CASE 48. *Pain in right iliac region and recurrent pelvic peritonitis dating from confinement three years ago; bursting of an abscess per* vaginam *eight months ago; persistent discharge of pus subsequently; acute symptoms during week preceding admission; mass behind and to right of uterus with physical signs of cellulitis and sinus in upper part of posterior vaginal wall; diagnosis of abscess connected with suppurative inflammation of right uterine appendages; abdominal section: pelvic contents matted; right ovary enlarged and containing numerous cysts, many of them filled with pus; fistulous communication between one of these and vagina; removal; recovery.*—A young married woman, aged 22, employed as a still-room maid, was admitted into St. Thomas's Hospital January 15th, 1891, on account of severe pain in the right iliac region and other symptoms.

Her marriage took place when she was eighteen. She was confined of her first and only child a year subsequently, and has never been quite well since. She had a greenish discharge for about two months after the labour, and suffered from time to time from pain in the right iliac region. This pain varied in intensity but never entirely disappeared, and twelve months ago she went into the Canterbury Hospital. She was there for a month and states that she underwent an operation of some kind. She remained well after this for three months, when the pain having returned, she one day whilst seated quietly in a chair, felt a sudden flow of discharge from the vagina. The discharge was thick, fœtid, yellow in colour, and very profuse. For two or three weeks the pain was easier, but it has never wholly disappeared. The discharge has continued with intervals to the present time, but since the first day has had no ill odour. A week ago she was suddenly seized in the night with acute pain in the right iliac region. The pain was relieved by poulticing, but the patient has since been quite unable to get about or resume her work.

The patient is in fairly good condition but anæmic. The temperature is normal.

On vaginal examination the uterus was found in normal position, the cervix was fixed by adhesions posteriorly. The pouch of Douglas was filled with a hard, rounded mass, extending further to the right side than to the left. The vaginal roof on the right side was slightly depressed. There was dense hardness in the tissues at the posterior vaginal reflection and immediately in front of the cervix; in the latter position simulating acute anteflexion of the uterus. At the upper part of the posterior vaginal wall was a small opening, the size of a pea, with indurated margins.

The diagnosis was chronic abscess in Douglas's pouch, communicating with the vagina, and connected with suppurative inflammation of the right uterine appendages.

Abdominal section was performed January 29th, 1891. The omentum was adherent to the pelvis. The pelvic viscera were densely matted by old adhesions; the broad ligaments were hard, rigid, and thickened. A loop of intestine and a band of omentum were adherent to the anterior abdominal wall just above Poupart's ligament on the right side. A soft, oblong mass was separated from its adhesions to the posterior aspect of the corpus uteri. This mass dipped down into Douglas's pouch, where its dense adhesions were separated with difficulty. The long axis of the mass was directed downwards. When fully separated and brought into view, it was found to be connected with the right broad ligament, and to consist of the much enlarged right ovary with the Fallopian tube stretched over and adherent to it. Both were removed. The appendages of the opposite side were then separated; during the process rupture of the ovary took place, a dark blood-clot escaping. The tube and ovary were removed, the greater part of the ovary remaining as part of the pedicle. The peritoneum was douched, and a glass drainage-tube introduced before closing the wound. A quantity of pus having flowed from the vagina during the operation, a vaginal douche of solution of corrosive sublimate, 1 in 5000 was administered. The operation lasted an hour and a half.

Description of parts removed.—The right ovary measured 2¼ inches by 1¾ inches by 1 inch. It consisted, on section, of a number of inflamed cysts, many of them full of pus, and all with hyperæmic walls. An opening, large enough to admit a goose-quill, and surrounded by granulation-tissue, was found on that part of the surface of the ovary which had lain most deeply in the pelvis. This opening communicated directly with one of the abscess-cavities in the substance of the ovary, and pus was seen exuding from it.

The right tube was attached to the ovary and was elongated. On section its lining membrane was found healthy and its canal empty.

The left Fallopian tube was beaded from kinking, but was otherwise healthy. No ovarian tissue was found in the parts removed on the left side.

The highest temperature recorded during the patient's convalescence was 99·8°. She was restless during the night of the 30th, and vomited several times. After this there was no vomiting. The drainage-tube was removed in forty-eight hours. Menstruation commenced February 1st and lasted five days. Some cystitis appeared on February 3rd but soon subsided under treatment. An abscess formed in the abdominal wall near the upper part of the wound, and burst on February 8th.

On February 27th a vaginal examination was made. There was a smooth, firm, tender swelling to the left of the uterus; none in Douglas's pouch or in the right side of the pelvis. There was a dimple in the post-vaginal wall at the site of the fistula.

At the beginning of March the patient again menstruated, and on March 7th she left the hospital well.

On November 15th, 1891, the patient was readmitted,

complaining of attacks of pain commencing in the right iliac region, lasting severely for a few hours and then gradually diminishing until they pass off in the course of about a week. She has had four such attacks; the first in June, the second in July, the third in September, and the last just before her readmission. There is vomiting during the first two days of each attack. The attacks have no connection with menstruation, which has been regular. Between the attacks the patient has felt well and strong. Temperature is normal. On vaginal examination no swelling could be detected on either side of the pelvis; the uterus was fairly moveable. There was a little tenderness on the right side.

This case exemplifies very strikingly the advantage of dealing with chronic abscess in the deeper part of the pelvis from above rather than from below. Had the treatment here consisted of enlarging the sinus in the posterior wall of the vagina and draining the abscess-cavity thus laid open, there would still have been numerous other abscesses to be reckoned with, that such an incision could not have reached. The opening found on the surface of the ovary was no doubt the aperture of communication with the vagina, due to ulceration of the wall of the cyst and of the parts to which it was adherent. The opening had been insufficient to allow of the complete emptying of the abscess; hence the persistent vaginal discharge. The absence of pyrexia before operation, notwithstanding the condition of the right ovary, is noteworthy, as also is the freedom from pelvic suppuration and sepsis during the recovery, considering that some soiling of the pelvis during the removal of the ovary must almost certainly have occurred.

The attacks of pain described by the patient as having occurred at intervals since the operation are probably to be explained by intestinal or omental adhesions at the site of operation.

CASE 49. *Pain in joints and high temperature for six*

weeks, regarded as due to acute rheumatism; discovery of purulent vaginal discharge; development of abdominal pain; patient found to be suffering from acute gonorrhœa; both sides of pelvis occupied by irregular swellings, right tube traced distinctly, enlarged and tortuous, left less distinct; abdominal section: pelvic contents matted; intraperitoneal abscess in Douglas's pouch fed by the openmouthed suppurating Fallopian tubes; removal of tubes and ovaries; recovery without suppuration; immediate disappearance of pyrexia and other pyæmic symptoms.— An unmarried girl, aged 24, a chambermaid at an hotel, was admitted into St. Thomas's Hospital, December 15th, 1890, under the care of Dr. Payne, for what appeared at first to be an attack of acute rheumatism. There had been pains in the right wrist for three days, and in the back of the neck, the left shoulder, left elbow, left leg and left knee for two days.

On admission the tongue was coated with a white fur; the temperature 102·2° to 102·8°; the pulse 120. There were coarse rhonchi heard over the upper part of the left lung and moist sounds near the apex of the right lung behind. The heart sounds were normal. The right wrist, left shoulder, left knee, and left tarso-metatarsal joints were tender and painful, without obvious effusion or any œdema or redness of the superjacent skin.

On December 22nd there was no pain or stiffness except in the left knee, which was stiff, swollen, and tender. The temperature has varied between 98·2° and 102°, the maximum record on the 16th having been 101·4°, on the 17th, 101·6°; on the 18th, 100·8°; on the 19th, 99·6°; on the 20th, 99·4°; on the 21st, 102°; and on the 22nd, 101·2°.

On January 15th, 1891, the temperature was 102·8°. The lungs were resonant everywhere, the breath-sounds normal; no unhealthy signs at apices; slight cough; no expectoration. Bowels regular. Tongue fairly clean. No tenderness about any joint. Left knee slightly swollen and kept in a position of flexion, extension causing pain.

On January 18th a vaginal discharge was noticed ; no abdominal pain ; temperature 99·4° to 104·4°.

On January 27th the patient having complained during the past three days of pain in the lower part of the abdomen with headache, sickness, and shivering, a suspicion, already existing, that the case was not one of ordinary rheumatism, was strengthened, and I was asked to see her and make a pelvic examination.

I reported that she was suffering from gonorrhœa and pelvic inflammation, and she was accordingly transferred, the same day, to Adelaide Ward, under my care.

The temperature since the last note had been as follows—(January 19th) 102° to 102·8°; (20th) 99° to 102·8°; (21st) 97° to 99°; (22nd) 97·6° to 102·2°; (23rd) 100·6° to 103·4°; (24th) 99·8° to 102°; (25th) 97° to 99·8°; (26th) 98·4° to 100·4°; (27th) 98° to 102·6°.

On being questioned with a view to determine if possible the date of infection, the patient stated that the only time she had been exposed to such a risk was on November 4th, 1890, when a stranger staying in the hotel took forcible advantage of her, and was, in consequence, dismissed from the hotel by the manager, to whom she reported the occurrence the same evening. During the five weeks she afterwards remained in her situation she had some pain on micturition and a vaginal discharge. She left her situation on December 11th. On awaking the following morning she for the first time felt pain in the right wrist. The remaining particulars of her illness have already been given.

On examination (after her removal to Adelaide Ward) there was found some pus on the vulva, and there was a copious flow of pus and mucus from the vagina on introducing the finger. There was slight redness at the posterior margin of the vaginal orifice ; no marked redness or swelling of the *meatus urinarius*, but pus issued from the *meatus* on making pressure along the urethra. There was no abnormal redness or swelling of the vaginal mucous membrane, or of the orifices of the ducts of Bartholin's glands. No pus exuded from the latter on pressure. Through the speculum some blood and mucus were seen issuing from the *os uteri*, on which was a broad ring of catarrhal erosion. Bimanually, there was felt in the right posterior quarter of the pelvis a firm resisting mass; and an elongated tube-like swelling could be felt passing outwards from the right cornu of the uterus, then turning downwards and backwards behind the uterus, forming a distinct cystic swelling in Douglas's pouch. Some thickening could also be felt in the left side of the pelvis, but of a less defined character. The uterus was inclined to the right.

The diagnosis was acute gonorrhœa, with pyosalpinx, pelvic peritonitis, and pyæmia.

Abdominal section having been proposed and agreed to, the operation was performed February 5th, 1891. The pelvic viscera were matted together by very firm adhesions behind and on each side of the uterus. On separating the tangled mass from the back of the *corpus uteri* some thick, inodorous pus made its escape, the finger passing into a cavity the size of a Tangerine orange. This cavity was the pouch of Douglas, walled in by the uterus, coils of intestine, and the uterine appendages.

The right appendages were now separated from their adhesions and brought into view. Closely adherent to them was the thickened vermiform appendix. On separating it the tip was ragged and bleeding; the distal end, to the extent of an inch, was therefore ligatured and removed. The tube and ovary were then removed. As there was some pus in the divided end of the tube in the stump, the stump was cauterised. The removed portion of the tube was thickened and full of pus, but showed no ulceration; its fimbriated end was widely open, allowing the contents to exude into the pelvic cavity. For some time the left appendages could not be found; they were

10

at length discovered, behind, adherent to, and wrapped round by the broad ligament. On removal the tube was found to be in a similar condition to its fellow on the opposite side, namely, thickened and full of pus, with the fimbriated end open, allowing the escape of its contents into the peritoneal cavity. Loops of thickened intestine were also adherent in the pelvis; for the most part they were left undisturbed.

The pelvis and abdominal cavity were well douched, a glass drainage-tube inserted, and the abdomen closed. The length of the incision was $2\frac{3}{4}$ inches. The duration of the operation, one hour and three quarters.

The patient made an excellent recovery. She was only once sick. The drainage-tube was removed in forty-eight hours. The temperature on the evening after the operation was 101·8°, after that it was generally normal and never reached 100°. There was no suppuration from the wound. Twelve days after the operation the patient was able to lie on the couch. She left the hospital well on the 28th March, the last three or four weeks having been devoted to treating the gonorrhœal inflammation of the cervix, vagina, &c.

The portion of right tube removed measured $3\frac{1}{2}$ inches in length and was bent at a right angle; its widest diameter (at the bend) was $\frac{3}{4}$ inch. The portion of left tube removed was $2\frac{3}{4}$ inches in length, and was also bent at a right angle, with a diameter of $\frac{3}{4}$ inch at the bend. Otherwise the left tube was smaller than the right.

This case may be commended to the consideration of those who disbelieve in the gonorrhœal origin of purulent salpingitis and general pelvic inflammation. The clinical evidence here is almost as complete as could be wished. The case is also an answer to those who recommend a long trial of rest and palliative treatment before operating. Nothing could have been gained here by waiting.

The manner in which the parts healed without a trace of suppuration, notwithstanding the extent of suppuration at the time of the operation and the prolonged manipulations that were required, is very noteworthy, as also is the rapid disappearance of the pyæmic symptoms.

CASE 50. *Pelvic pain with diarrhœa and hæmorrhage from bowels, alternating with constipation for six weeks, attributed to getting wet; admission to medical wards as a case of typhoid; on vaginal examination an irregular, long, smooth swelling found in left side of pelvis with some indistinct thickening on right; history of impure connection and vaginal discharge; diagnosis of left pyosalpinx; abdominal section: left pyosalpinx communicating by an ulcerated opening with a suppurating ovarian cyst; right tube thickened and occluded; both tubes and ovaries removed; rapid recovery.*—A servant girl, aged 19, single, was admitted into the medical wards of St. Thomas's Hospital February 10th, 1881, supposed to be suffering from enteric fever.

She had been wet through on the 3rd of January, and during the night had been seized with severe pain in the right iliac region. For a week she tried to get through part of her work, but from that time had been obliged to be in bed. Three weeks before admission there was noticed a considerable quantity of blood in the motions on two successive days. She was at that time suffering from diarrhœa. For the fortnight before admission the bowels were constipated. On the Friday and Monday before admission there had again been blood in the motions, but less in quantity. During the whole of the past six weeks there have been headache, loss of flesh, and pains in the limbs. The patient has also had a yellow discharge from the vagina.

On February 18th she complained of a good deal of pain in the lower part of the abdomen, thighs, and back, and lay on her back with the knees drawn up. The temperature had varied since admission from normal to 100·8°.

I was asked to see and examine her the following day. The uterus was retroverted and its mobility impaired.

There was an irregular but somewhat elongated and smooth swelling in the left posterior quarter of the pelvis, and some less distinct thickening in the right. I found on enquiry that the girl had frequently had sexual intercourse between June, 1889, and the middle of 1890, but that nothing of the kind had taken place after the latter date until a week before the commencement of her present illness. I gave it as my opinion that the patient was suffering from pyosalpinx on the left side and some thickening of the right tube, with secondary peritonitis, the disease being either gonorrhœal or tubercular. She was thereupon transferred to Adelaide Ward on February 21st. On the 24th an examination was made under ether, with the result of confirming the opinion already given.

Abdominal section was performed on February 26th.

In the left posterior quarter of pelvis was found a thin-walled, not very tense, soft, cystic swelling, with tube attached to, if not forming part of it. The tumour was easily separated, the adhesions, though universal, being slight in character and recent. Notwithstanding the gentlest handling, the cyst-wall gave way and a purulent discharge welled up. On bringing the mass into view it was found to be a suppurating cyst of the left ovary, communicating by an ulcerated opening the size of a pea with the Fallopian tube, which was thickened and contained pus amongst its inflamed rugæ. The right tube was enlarged, occluded, and adherent; the right ovary was normal. Both tubes and both ovaries were removed. The peritoneal cavity was douched, a drainage-tube inserted, and the abdominal wound closed. There was a good deal of oozing from separated adhesions, the arresting of which occupied a good deal of time, and the operation lasted an hour and a half.

The patient made a rapid recovery and left the hospital well on April 1st. The suture-tracks in the abdominal wound suppurated, which is an unusual occurrence, but there was no purulent discharge from the pelvis. A metrostaxis commenced on the day of operation and lasted until March 5th, after which there was an offensive vaginal discharge for several days.

The patient was sent to a convalescent home; but she was dismissed from there for bad conduct, and I have heard nothing of her since.

The case is of special interest as showing the communication between tube and ovary in actual process of formation. There was probably ulcerative salpingitis in the first instance with adhesion to a cystic ovary, followed by perforation of tube and cyst-wall and infection of contents of the cyst.

PART III.

No classification of such a series of cases as that here recorded can be altogether free from objection; but the following table will, I trust, be found fairly satisfactory. In order to prevent unnecessary repetition I may premise that in all the cases but one there was marked pelvic peritonitis. The exception was Case 32, in which I made a wrong diagnosis, mistaking for inflamed and adherent appendages a retroflexed uterus, enlarged and distorted from fibroids, and incarcerated beneath the sacro-vertebral promontory. I have included the case here because admission to this series has been determined by the object for which the operation was undertaken, and not by what was found. Setting this case aside then for the present, the conditions causing or associated with the peritonitis in the remaining 49 cases were as follows:

Tubercular disease of Fallopian tube (Case 21 and 29) . 2
Suppurating salpingitis (Cases 7, 9, 14, 15, 16, 17, 18, 20, 25,
   27, 28, 30, 33, 36, 37, 40, 43, 45, 49, 50) . . 20
Non-suppurating salpingitis, including six cases complicated
   with suppurating ovarian cyst (Cases 1, 2, 4, 12, 19, 24,
   26, 35, 39, 41, 46, 48) 12
Pelvic abscess, seat undetermined (Cases 5, 6, 13) . . 3
Suppurating, pedunculated, retro-peritoneal cyst (Case 10) . 1

Abscess in abdominal wall (? tubercular) with masses of enlarged pelvic glands and miliary tubercle of peritoneum (Case 42)   1
Hæmatocele (Cases 23 and 31) 2
Hæmatosalpinx with hæmatocele (Cases 34, 44, and 47) . 3
Hæmatoma of broad ligament (Case 22) . . . 1
Broad ligament cysts—
   (a) With ovaritis (Cases 1 and 38) . . 2
   (b) With hydrosalpinx (Case 3) . . 1
                                      —   3
Encysted serous effusion (Case 8) . . . 1
                                              —
                                            49

The cases of suppurating salpingitis may be subdivided as follows :

(a) With occlusion (pyosalpinx) (Cases 7, 15, 30, 40, 43) . 5
(b) With distal end open (Cases 16 and 36) . . 2
(c) With suppurative disease of the ovary (Case 37) . . 1
(d) With a direct communication between the tube and a suppurating cyst of the adjacent ovary (suppurating tubo-ovarian cyst) (Cases 17, 18, 20, 25, 33, 50) . . . 6
(e) With non-suppurating cystic ovary (Case 27) . . 1
(f) With suppurating hæmatocele (Case 14) . . 1
(g) With hydrosalpinx (Cases 9 and 45) . . 2
(h) With intra-peritoneal abscess (Cases 28 and 49) . . 2
                                            —
                                            20

The cases of non-suppurating salpingitis may be classified into—

(a) Uncomplicated cases (Cases 19 and 24) . . . 2
(b) With suppurating ovarian cyst (Cases 4, 12, 26, 39, 41, 48) . 6
(c) With non-suppurating ovarian cyst (Cases 35 and 46) . 2
(d) With hæmatosalpinx and hæmorrhagic ovarian cyst (Case 2) 1
(e) With double hæmatocele (Case 11) . . . 1
                                            —
                                            12

Number of cases in which there was pelvic suppuration.—Perhaps the most interesting point brought out, on analysing these cases, is the large proportion in which there was some form of pelvic suppuration. Thus, out of the total number of fifty, this condition existed in no fewer than thirty, *i. e.*, in 60 per cent. With regard to the seat of the suppuration, in thirteen cases it was the Fallopian tube alone; in six cases it was the ovary alone; while in seven cases it was both tube and ovary, the two being, in six of these, in direct communication. In the remaining four cases the seat of suppuration was either not accurately determined, or, as in Case 10, did not involve either tube or ovary. In no instance was there evidence of the suppuration being in the pelvic connective tissue.

Origin of the suppuration.—I hope at some future time to discuss more fully than is here possible, the etiology of suppurative inflammation of the uterine appendages. In the meantime I may say that, the larger my experience, the less disposed I am to attribute to catarrh anything like the share it is popularly supposed to have, in causing pelvic inflammation. Even cases like Nos. 27 and 36, where the evidence in favour of a catarrhal origin seems at first sight indisputable, prove on further investigation to be chronic cases, in which exposure has merely had the effect of producing an acute exacerbation. The real causes of pelvic inflammation in the great majority of cases will, I believe, eventually prove to be sepsis, gonorrhœa, and perhaps tubercle. Amongst the cases here recorded, the evidence of gonorrhœal origin is very strong in a good many cases, and in at least five cases (Nos. 9, 14, 15, 43 and 49) seems irresistible.

Mortality.—The total number of fatal cases was nine, a mortality of 18 per cent. The cause of death in four cases (3, 9, 10, 16) was peritonitis, no doubt septic; in one case (11) the only lesion discovered at the autopsy was acute nephritis; in another case the patient had intestinal obstruction; an artificial anus was made, and death occurred next day from peritonitis. I have little doubt that the obstruction was really due to septic peritonitis. In the remaining three cases no *post-mortem* examination was made. One of the patients (38) died suddenly from collapse on the eleventh day; the other died with symptoms of septic peritonitis. Of the patients who died, one was a case of tubercular disease of the Fallopian tubes; two were cases of purulent salpingitis; two were cases of suppurating tubo-ovarian cyst; two were cases of very chronic pelvic peritonitis, in which very little was removed at the operation; one was a case of double salpingitis, non-purulent, with a small hæmatocele at the open mouth of each tube; and one was a case of hæmorrhagic retro-peritoneal cyst, with abscesses in its walls.

Nature of operation.—The operation involved the complete removal of the appendages in 16 cases, and their partial removal in 23. In the remaining 11 cases none of the appendages was removed. Of the 16 complete removals, 15 recovered; of the 23 partial removals, 17 recovered; of the 11 patients in whom neither tube nor ovary was removed, 9 recovered.

Flushing of peritoneum.—The peritoneal cavity was flushed with hot solution of boric acid in 22 cases, 18 of which recovered.

Drainage.—The drainage-tube was used in 47 out of the 50 cases. In 38 cases, the glass drainage-tube alone was employed; the length of time it was kept in was as follows :—Twenty-four hours in 14 cases; thirty-six hours in 4 cases; forty-eight hours in 14 cases; sixty hours in 4 cases; seventy-two hours in 2 cases.

In 7 cases an india-rubber tube was substituted for the glass tube; at the end of twenty-four hours in 1 case, forty-eight hours in 4 cases, and seventy-two hours in 2 cases.

In 2 cases india-rubber tubes were employed throughout.

Fæcal fistula.—In 2 cases, a fæcal fistula formed after the operation; spontaneous closure took place in each instance.

Pain.—In the large majority of the cases pain was permanently relieved. Almost all the patients who recovered have returned to the hospital to report themselves at more or less prolonged intervals after their discharge. Only five of these have complained of pelvic pain.

Sinus at lower angle of wound.—In 6 cases it is noted that a sinus existed when the patient went home; in only 2 of these has healing failed to take place since (Nos. 7 and 41).

Hernia at site of abdominal wound.—Four patients have developed a hernia at the line of incision. One of them had had the abdomen opened twice.

On the whole, the rapidity of convalescence and freedom from unpleasant sequelæ have been remarkable. Of the forty-one patients who recovered, twenty-four escaped without the slightest suppuration (except in one or two instances in the suture-tracks), including no fewer than nine cases of suppurative disease of the tubes, two cases of suppurating ovarian cyst with salpingitis, and two cases of suppurating tubo-ovarian cyst.

Hæmorrhage as a symptom of tubal inflammation.—The effect of tubal inflammation upon the menstrual function is illustrated by the following figures. Out of the thirty-two cases of salpingitis included in the present series, twelve had more or less continuous hæmorrhage, five had amenorrhœa, three had dysmenorrhœa, and twelve menstruated normally. Dividing the cases into purulent and non-purulent salpingitis, we find that amongst twenty cases of purulent salpingitis, eight had metrorrhagia, three had amenorrhœa, three had dysmenorrhœa, and six had no disturbance of menstruation. Of the twelve cases of non-purulent salpingitis, complicated and uncomplicated, four had metrorrhagia, two had amenorrhœa, whilst in six there was no interference with the menstrual function. So far, therefore, as the small number of cases here recorded enables us to judge, irregular uterine hæmorrhage is a symptom of salpingitis in rather more than a third of the cases, or to speak more precisely, in two-fifths of the purulent cases, and in one-third of the non-purulent. The hæmorrhage is seldom profuse, and appears never to be in itself a source of danger.

The temperature as a guide to the diagnosis of pelvic suppuration.—It is generally held that if the temperature is not raised, it is a fairly certain indication that there is no suppuration. The following figures show that this test is unreliable. In twelve of the thirty cases in which suppuration was present the temperature before operation was absolutely normal. In one case there was a single rise of temperature (after examination) to 103°; in another case the temperature only twice exceeded the normal during a period of six weeks; in a third case there was but a single rise of temperature in ten days, and that only to 100°; in a fourth case, during a period of eight days, the temperature only on one occasion exceeded 100°, and in a fifth case the highest record was 100·4°. In twelve cases the temperature was distinctly febrile. Of one case I have no note of the temperature before operation. A much more valuable guide to the diagnosis of the presence of pus in the pelvis is the recurrence, on comparatively slight provocation or without ostensible provocation of any kind, of more or less severe attacks of pelvic peritonitis, after apparent recovery from the first attack. What happens in such cases is that the pus becomes enclosed, and for a time gives no sign of its presence. Then comes some slight exciting cause, and the purulent collection becomes the centre of an acute and wide-spread inflammation. Or, in the absence of such exciting cause, the tissues enclosing the pus undergo ulceration, until at last perforation occurs, and the pus, after having been imprisoned, it may be for months or years, is set free in the pelvis or escapes into some neighbouring viscus or canal. I do not propose in this paper to enter into an elaborate defence of the operation of which it treats. My object is to present a statement of facts, and to let them speak for themselves. The operations here described were not "done in a corner." With few exceptions they were performed at St. Thomas's Hospital before the resident officers and students, and any colleagues or other visitors who cared to witness them. Being a new departure, they were watched with keen interest. The parts removed were submitted, while still fresh, to the curator of the hospital museum, Mr. Shattock, who examined them then and there, and is responsible for the description recorded in the notes. It is scarcely possible to have more complete guarantees against reckless surgery or inaccuracy of statement. It would be absurd to maintain that every case in such a long series was a suitable one for operation; but the instances in which I had reason to regret having operated were exceedingly few, and were much less numerous than those in which I regretted not having operated sooner. In the remarks appended to the individual cases, I have endeavoured honestly to confess my mistakes. The operations here recorded have nothing in common with those which are undertaken merely for the relief of pelvic pain without obvious lesion. Of these latter I have no experience. The only instances in which I have removed the normal tubes and ovaries are those in which the operation has been performed for uterine fibroids. I make this statement in order to limit any discussion that may follow the reading of this paper, to the operation with which it deals, an operation which I believe to be founded on sound surgical principles, and destined to take its place amongst the established operations of modern surgery.

I shall be disappointed, however, if this communication is regarded merely as a plea for more frequent recourse to surgical treatment. I trust it may also have some value as a contribution to our knowledge of the diagnosis and pathology of some of the most common diseases of the female pelvic organs, especially tubal disease and its numerous and very serious complications.

Postscript (August 15th, 1892).—During the eighteen months that have elapsed since February, 1891, when the above paper was commenced, I have completed a second series of fifty operations of a similar kind. In this series the mortality has been less than half that of the first fifty, nine deaths having occurred in the first fifty, and four in the second. Amongst the last twenty-five cases operated upon, not one has proved fatal. It is therefore, I think, fair to say that the mortality in this, as in most serious operations, tends to diminish with increased experience. Of the four patients in whom the operation proved fatal, at least three would have died within a very short time if they had not been operated upon; the operation was too late to save them.

The fatal cases included—

    1 suppurating ovarian cyst.
    1 suppurating tubo-ovarian cyst.
    1 tubercle of ovary.
    1 pelvic abscess of uncertain origin.

The details of the second fifty cases cannot of course be given without unduly prolonging an already too long paper. It may be interesting, however, to append a classified list of them.

Tubercular disease of Fallopian tube	1
Suppurating salpingitis (including two cases of suppurating ovarian cyst, and two of suppurating tubo-ovarian cyst)	20
Non-suppurating salpingitis (complicated in four cases by suppurating ovarian cyst)	6
Suppurating subperitoneal cyst	1
Pelvic abscess, seat uncertain	3
Tubercular disease of ovary, with suppuration	3
Suppurating ovarian cyst (complicated in one instance by inflammation of the vermiform appendix)	4
Hydrosalpinx	1
Serous cyst of ovarian ligament	1
Inflamed ovarian cyst	1
Dermoid cyst of ovary	1
Perityphlitis, after delivery, with suppuration	1
Tubercular peritonitis	1
Malignant disease of pelvis	1
Hæmatosalpinx	4
Unruptured tubal gestation, with apoplectic ovum	1
	50

The specimens from eleven of these cases have been exhibited to this Society, and descriptions, accompanied with a brief clinical history, have been printed in its 'Transactions.'* Six other of the cases have been published in detail in the 'Lancet.'†

* A series of seven cases of Pyosalpinx, shown November 4th, 1891. Two cases of Tubal Gestation with Apoplectic Ovum, shown May and June, 1892. Two cases of Pyosalpinx, shown July, 1892.

† See "Mirror of Hospital Practice" in 'Lancet' for July 2nd and 9th, 1892. "Six cases of Abdominal Section for Recurrent Pelvic Peritonitis."

No.	Name.	Occupation and residence.	Age and civil condition.	Date of operation.	Place of operation.	Symptoms.	Duration of illness.
1	A. McC.	Dressmaker, Manchester	25 W.	1885 May 13	St. Mary's Hospital, Manchester	Anæmia; emaciation; constant pain in left iliac region; inability to sit, and hence to follow occupation	6 years
2	M. M.	Winder in cotton mill	26 M.	Oct. 7	"	Anæmia; emaciation; continuous pain in lower part of abdomen, especially on right side; metrorrhagia (two months)	7 years
3	J. R.	Housekeeper, Manchester	35 M.	1886 Jan. 13	"	Continuous pain in pelvis; repeated attacks of pelvic peritonitis	10 years
4	M. B.	House work, Royton	25 M.	April 30	"	Recurrent pelvic peritonitis; constant pelvic pain, incapacitating her for work	3 years
5	M. E. B.	Weaver, Rawtenstall	21 S.	1887 June 7	"	Amenorrhœa 19 weeks, severe pain in lower part of abdomen, commencing with acute attack 10 weeks before admission. After 2 months' rest in hospital pain and tenderness subsided, but swelling increased	19 weeks

Physical signs and diagnosis.	Condition found and nature of operation.	Glass drainage-tube.	Peritoneum flushed.	Result.	Remarks.
Firm, rounded, tender swelling to right of and behind uterus; uterine mobility impaired. *Diagnosis.*—Chronic ovaritis with pelvic peritonitis	Right ovary size of walnut, inflamed and indurated. Firm tumour of each broad ligament, consisting of a compact mass of small cysts; left ovary healthy; all parts adherent. Tumours enucleated; right ovary and tube removed	48 hours	No	R.	Seven months later stout and well, free from pain, and able to earn her living. See 'Brit. Med. Journ.,' Jan. 30, 1886.
Oblong, firm, tender swelling on right side of pelvis, pushing uterus to left. Probably the right Fallopian tube distended	Fallopian tube on right distended with blood; its walls thickened; right ovary enlarged to size of hen's egg, containing a bloodcyst; left ovary cystic. Both ovaries and right tube removed; adherent viscera separated	72 hours	No	R.	Temp. during convalescence only once reached 100° F. Two months after operation free from pain, able to go about as usual. See 'Brit. Med. Journ.,' Jan. 30, 1886.
Uterus fixed; tender swelling in right side of pelvis. *Diagnosis.*—Chronic ovaritis with adhesions	Contents of pelvis matted; right tube distended with serum; three serous cysts in broad ligament. Cysts and diseased tube removed; adhesions separated	Until death	No	D.	Died on third day from peritonitis.
Small fixed tumour on right side of pelvis, size of orange. *Diagnosis.*—Dilated right tube	Right ovary cystic and enlarged, 3 in. long; one large cyst filled with pus; universally adherent; both tubes much thickened, with cysts in walls. Both tubes and both ovaries removed	48 hours	No	R.	Convalescence rapid. Six months after operation stout, well, and free from pain. Oct. 25, 1892.—Feels as well as ever she was; no pelvic pain; has not menstruated for two years.
Abdomen swollen, tender, resonant; no tumour; no fluctuation; uterus normal in size, fixed. *Diagnosis.*—Uncertain	All contents of pelvis matted together by adhesions; tense abscess on right side of pelvis, with thin walls. Abscess emptied, irrigated, and drained; walls secured to abdominal incision. Uterus and appendages not distinguished	24 hours. India-rubber tube 5 months	No	R.	Convalescence rapid, except that a discharging sinus existed for many months. In August, 1892, she was perfectly well, and had been married 2 years. Menstruated regularly.

No.	Name	Occupation and residence.	Age and civil condition.	Date of operation.	Place of operation.	Symptoms.	Duration of illness
6	E. J.	None, Winton, near Patricroft	23 M.	1887 Oct. 12	St. Mary's Hospital, Manchester	Menorrhagia; abdominal pain dating from two months after marriage, and becoming worse; incapable of least exertion	13 mo.
7	E. F.	Servant, Ashford	25 S.	1888 April 5	St. Thomas's Hospital, London	Amenorrhœa; emaciation; pyrexia; constant pain; bedridden	16 months
8	A. L.	None, Southwark	20 W.	May 21	,,	Pain in left iliac region and in micturition. Pallor, emaciation, general feeling of illness. Temp. 102·6°	12 mos
9	M. C.	Tailoress, Dalston	19 S.	Oct. 18	,,	Gonorrhœa; recurrent pelvic peritonitis; constant pelvic pain	9 mont
10	S. T.	Dressmaker, Marylebone	32 S.	Dec. 20	,,	Recurrent pelvic peritonitis; almost constant pelvic pain, especially on left side. Looks thin, sallow, ill, and tired	5 yea
11	G. C.	House work, Barking	32 M.	1889 Feb. 21	,,	Hæmorrhage; pain in back, vulva, right thigh, and knee	13 week

Physical signs and diagnosis.	Condition found and nature of operation.	Glass drainage-tube.	Peritoneum flushed.	Result.	Remarks.
Large fluctuating swelling to right of and behind uterus, which is normal in size and pushed forwards. *Diagnosis.*—Retro-uterine abscess	Intra-peritoneal abscess in right side of pelvis; 20 fl. oz. pus removed; cavity irrigated and drained; edges secured to abdominal incision; nothing removed	72 hours. India-rubber tube 5 months	No	R.	Rapid improvement in general health, but discharging sinus existed for several months. In July, 1892, was quite well.
Tense, hard, obscurely fluctuating tumour in left iliac region; uterus fixed. *Diagnosis.*—Pelvic abscess	Abscess to left of uterus; 3 fl. oz. pus removed; wall ¼ in. thick, lined with caseous material. Cavity emptied and drained; opening secured to abdominal incision	India-rubber tube many months		Yes	R.	Immediate improvement in health, but sinus left, discharging muco-pus. On Jan.14, 1890, sinus dissected out; found to consist of left Fallopian tube, thickened, but no longer distended.
Tense, fluctuating, slightly prominent swelling above pubes; uterus fixed and displaced to right; brawny swelling above vaginal roof on left side. *Diagnosis.*—Pelvic abscess	Intra-peritoneal effusion of serum (20 fl. oz.) walled in by pelvic viscera and by adhesions. Cavity emptied and drained	India-rubber tube 11 days	No	R.	Highest temp. after operation, 99° F. Remained well and at work for 4½ years. In Sept., 1892, attack of pelvic pain; mass on right side of uterus; right appendages removed for chronic inflammatory disease.
Tense, fixed, cystic swelling behind and to right of uterus; uterus fixed; thickening on left. *Diagnosis.*—Right hydrosalpinx	Right hydrosalpinx. Tube and adjacent ovary removed	No	No	D.	P.M.—General peritonitis; pus in left tube and in remains of right tube. See 'Obst. Soc. Trans.,' vol. xxx, p.406 and plate; also 'Brit. Med. Journ.,' July 20, 1889, pp. 123, 124.
Indistinct, globular, fluctuating tumour above pubes, causing little or no prominence of abdominal wall. Uterus pushed to left	Pelvic viscera densely matted; retro-uterine, pedunculated, subperitoneal cyst, containing 30 fl. oz. dark brown fluid (altered blood) and two small suppurating cysts in its wall	20 hours	Yes	D.	P.M.—Suppurative peritonitis. Both ovaries and both tubes involved amongst the pelvic adhesions. See 'St. Thomas's Hosp. Rep.,' vol. xviii, p. 76.
Smooth, firm, elastic, immoveable mass behind and to right of uterus; obscure thickening high up in left posterior quarter of pelvis	Old pelvic adhesions; amongst them on each side a firm blood-clot embraced by the fimbriæ of the Fallopian tube; right tube thickened, empty, and undilated. Both tubes and left ovary removed with the clots	50 hours	No	D.	Died on ninth day. P.M.—Acute nephritis. No cause of death discovered in parts concerned in the operation. See 'St. Thomas's Hosp. Rep.,' vol. xix, p. 179.

11

No.	Name.	Occupation and residence.	Age and civil condition.	Date of operation.	Place of operation.	Symptoms.	Duration of illness.
12	S. A. W.	Servant, Worfield, Bridgnorth	22 S. 1-para	1889 March 21	St. Thomas's Hospital, London	Recurrent pelvic peritonitis; constant pain right iliac region; inability to work or even move about; pyrexia	5 years
13	A. M.	None, Newington	39 M.	Aug. 2 and Aug. 30	,,	Constant pelvic pain; purulent discharge from rectum; emaciation; anæmia	2 years
14	A. O.	Dressmaker, Waterloo Road	28 M.	Sept. 14	,,	Pain, loss of appetite, great emaciation; pyrexia; inability to sit, and therefore to earn living	12 months
15	S. B.	Prostitute, London	22 M. (?)	Oct. 17	,,	Recurrent pelvic peritonitis; irregular hæmorrhage. Purulent discharge from uterus	4 months
16	L. B.	Nursemaid, Turner's Hill	34 S.	Oct. 24	,,	Dysmenorrhœa; pain in right hip and left iliac region, the pain latterly constant; always ailing	11 years
17	A. C.	None, Edmonton	40 M.	Nov. 18	,,	Seized 5 weeks before admission with stabbing pain in lower part of abdomen. Subsequently had general bronchitis, pains of rheumatic character, abdominal pain and vomiting, with temp. 102°. On admission extremely ill; bronchopneumonia, occasional vomiting, much pain in right iliac region and down right leg. Temp., Nov. 8th to 18th, varied from 99·8° to 104·8°	5 weeks

Physical signs and diagnosis.	Condition found and nature of operation.	Glass drainage-tube.	Peritoneum flushed.	Result.	Remarks.
Uterus pushed to right by a smooth, firm, elastic, slighty moveable mass, filling up left posterior quarter of pelvis	Small suppurating cyst of left ovary; both tubes thickened and dilated; right ovary twice normal size and adherent. Both ovaries and both tubes removed	20 hours	No	R.	Temp. at no time exceeded 100° during convalescence. Apr.25. —Sent to Convalescent Hospital, Eastbourne. Jan., 1891.—Quite well and at work as a domestic servant. See 'St. Thomas's Hosp. Rep.,' vol. xix, p. 155.
Hard mass behind and to left of uterus. *Diagnosis.*— Pelvic abscess	Thick-walled abscess deep in left side of pelvis; 1½ fl. oz. pus withdrawn; edges secured to margins of abdominal incision	20 hours	No	R.	No pus from rectum after operation. Jan., 1891. — Ventral hernia, otherwise quite well.
Tender, irregular swelling behind and to right of uterus, displacing uterus to left. *Diagnosis.*—Pyosalpinx	Purulent salpingitis with suppurating hæmatocele. Left tube removed	48 hours. India-rubber tube 2 weeks	Yes	R.	Acute pneumonia during convalescence. Health restored by operation. See 'Brit. Med. Journ.,' Dec. 27, 1890.
Fixed, ill-defined, irregular mass in right posterior quarter of pelvis. *Diagnosis.*—Pyosalpinx	Right tube occluded, filled with pus; left to external appearance normal. Right tube removed	24 hours	No	R.	Rapid recovery. Temp. uniformly normal. A month later, uterus curetted, &c. March 7th, 1891.—Quite well, menstruation regular, no discharge, condition of pelvis normal.
Uterus fixed; hard, irregular mass behind uterus, connected with sausage-shaped swelling traceable to right cornu of uterus; tenderness in left posterior quarter of pelvis; no swelling. *Diagnosis.*—Disease of right Fallopian tube	Right tube enlarged and adherent. Left tube apparently normal. Right tube and ovary removed	48 hours	Yes	D.	P.M.—Pus found in uterus and in *left* tube.
Ill-defined, soft, elastic swelling in lower part of abdomen, extending from right lateral wall of pelvis nearly to left, appreciable *per vaginam*, where it is smooth, uniform, tense, and elastic. Uterus fixed, pushed forwards and to left. *Diagnosis.* —Pelvic suppuration; septicæmia	Tumour aspirated, 18 fl. oz. fetid pus withdrawn. Operation two days later. Right tube much elongated and enlarged, with thickened walls, communicating with ovarian cyst by opening large enough to admit finger, contents suppurating. Left ovary cystic, size of orange, inner surface papillomatous. General adhesions. Both ovaries and both tubes removed	44 hours	Yes	R.	Broncho-pneumonia (septic?) at time of operation. Temperature, evening of operation, 101·6°; afterwards never exceeded 99·6°. July 17, 1891. —Remains well. See 'St. Thomas's Hosp. Rep.,' vol. xix, p. 165.

No.	Name.	Occupation and residence.	Age and civil condition.	Date of operation.	Place of operation.	Symptoms.	Duration of illness.
18	C. D.	Norwood	29 M.	1889 Nov. 25	St. Thomas's Hospital, London	Recurrent pelvic peritonitis. Acute pain in right side of abdomen; hæmorrhage. On admission, pale, thin, and extremely ill; temp. 102·6°; resp. 40; sordes on teeth and lips	7 years
19	A. H.	None, Clapham	27 M.	Nov. 28	,,	Pain in left iliac region; pyrexia	2½ years
20	R. H.	None, Wandsworth	54 M.	Dec. 5	,,	Weakness; pallor; hæmorrhage; temp. normal in morning, 99·8° to 100·4° in evening; dull pain in lower part of abdomen; swelling of legs and feet	6 or 8 weeks
21	Mrs. C.	None, Luton	30 M.	Dec. 24	St. Thomas's Home	Severe paroxysmal pain lower part of abdomen and back; menorrhagia; night-sweats; emaciation	5 months

CERTAIN CASES OF PELVIC PERITONITIS.

Physical signs and diagnosis.	Condition found and nature of operation.	Glass drainage-tube.	Peritoneum flushed.	Result.	Remarks.
Irregular swelling felt deeply in lower part of abdomen; *per vaginam* tense, smooth, elastic, swelling filling up right side of pelvis; uterus to right and fixed. *Diagnosis.*—Pelvic suppuration; septicæmia	Right tube much thickened and lengthened, communicating with cyst of ovary by opening ¼ in. in diameter, contents suppurating. Left tube also in a state of suppurative inflammation. Left ovary not seen. General adhesions. Both tubes and ovarian cyst removed	50 hours	Yes	D.	Died from peritonitis, 5.30 a.m., Nov. 29, having had artificial anus made previous day for intestinal obstruction. See 'St. Thomas's Hosp. Rep.,' vol. xix, p. 168.
Uterus fixed; irregular, hard mass passing outwards from each cornu, that on left passing forwards, that on right backwards. *Diagnosis.* — Double salpingitis	Both tubes thickened, occluded, and densely adherent; ovaries adherent. Ovaries and tubes removed	24 hours	No	R.	See 'Brit. Med. Journ.,' Dec. 27, 1890. Jan., 1891. — Quite well, except for a small ventral hernia.
Rounded, firm, smooth, lobulated tumour above pubes on left; a swelling on right less firm, with tense band of tissue running transversely across it. Tumour on left is uterus enlarged; that on right separate from it. *Diagnosis.* — Fibroid enlargements of uterus; ovarian cyst behind right broad ligament	Right tube irregularly distended, communicating with cyst of ovary by aperture large enough to admit a goose-quill. Portion of tube removed 6½ in. long; contents suppurating, fetid. Left ovary cystic, 1¼ in. × ⅜ in., removed to check growth of bleeding fibroid	26 hours	Yes	R.	Acute endocarditis during convalescence. See 'St. Thomas's Hosp. Rep.,' vol. xix, p. 172.
Uterine mobility impaired; high up on right side elongated swelling, tender and tortuous. *Diagnosis.* —Chronic inflammation of right tube and pelvic peritoneum, probably tubercular	Chronic inflammation of both tubes; cystic disease of right ovary; dense peritoneal adhesions; miliary tubercles on peritoneum of tubes, intestine, and uterus. Tubercular ulcers in both Fallopian tubes, filled with caseous matter	24 hours	Yes	D.	See 'Brit. Med. Journ.,' Dec. 27, 1890.

No.	Name.	Occupation and residence.	Age and civil condition.	Date of operation.	Place of operation.	Symptoms.	Duration of illness.
22	L. T.	None, Battersea	27 M.	1890 Jan. 17	St. Thomas's Hospital	Pallor; anxiety of countenance; severe pain in left iliac region; high temperature	2 weeks
23	K. A.	None, Kentish Town	28 M.	Jan. 21	,,	Attack ushered in by pain and vomiting; 9 weeks after last menstruation; since that continuous hæmorrhage	3½ months
24	E. B.	Cricketer, King's Cross	18 S.	April 10	,,	Recurrent pain in left iliac region; vomiting; pyrexia. History of yellow vaginal discharge for two years	Acute symptoms 6 weeks
25	Mrs. L.	Stationer, Slough	38 M.	May 19	St. Thomas's Home	Pain in left side since miscarriage 6 years ago. Was taken acutely ill, August, 1889, at Margate, after getting wet, and has been in bed almost ever since with abdominal pain. Occasional offensive discharges of matter from rectum. Temp. 100° to 103° until March; since March normal	6 years; acute symptoms 9 months
26	M. J. H.	Laundress, Tooting	25 M.	May 22	St. Thomas's Hospital	Not well since miscarriage 12 months ago; lost flesh and had pain in left iliac region; pain worse during and since last period, with difficulty of micturition and pain before defecation	Acute symptoms 6 weeks

CERTAIN CASES OF PELVIC PERITONITIS.

Physical signs and diagnosis.	Condition found and nature of operation.	Glass drainage-tube.	Peritoneum flushed.	Result.	Remarks.
Large, tense, tender mass in left posterior quarter of pelvis and behind uterus, pushing uterus to right. *Diagnosis.* — Pelvic abscess	Exudation in left broad ligament with even surface, and soft but firm consistence; adhesive peritonitis; appendages normal. Probably a hæmatoma. Nothing removed	7 hours	No	R.	On Feb. 7 mass much less in all dimensions; temp. normal.
Oval swelling size of orange behind uterus and left broad ligament. *Diagnosis.*—Uncertain	Old intra-peritoneal blood effusion. No organised structure discovered. Tubes and ovaries adherent, but presenting no marked lesion; not removed	48 hours	Yes	R.	Probably a so-called tubal abortion.
Tense, elongated, non-fluctuating, fixed swelling in left posterior quarter of pelvis, with small, firm body enclosed in its fold. *Diagnosis.*—Inflamed left tube, enclosing normal ovary; both adherent	Both tubes thickened from old inflammation; mucous membrane healthy; no fluid in canal; right tube thicker than left; both firmly adherent. Ovaries healthy, adherent. Tubes and ovaries removed	20 hours	No	R.	Much vomiting and pain up to April 27, with alarming emaciation; after which recovery rapid. March 5, 1892. — Has had no pain since leaving hospital. Has menstruated regularly. Is well and strong.
Considerable abdominal swelling with hardness and resistance over left side and rounded prominence in middle line. Cervix uteri pushed upwards and forwards; large fluctuating swelling behind, depressing retro-uterine pouch. *Diagnosis.*—Pelvic suppuration, probably of ovarian cyst, fistulous opening into rectum	Both Fallopian tubes thickened and elongated, stretched over large suppurating ovarian cysts, with which the tubes were in direct communication by openings, that on the left large enough to admit little finger, that on right smaller. Both cysts removed with the tubes	72 hours	Yes	R.	Convalescence protracted. In February, 1891, presented herself, looking stout and well. Sinus still discharging slightly; no swelling in either posterior quarter of pelvis; menstruated four times in 5 months; duration normal, quantity variable. No pain, but back aches after exertion. Aug., 1892. —Quite well; sinus healed 8 months ago.
Tumour in left iliac region, felt but not seen; uterus pushed to right, fixed; tense, tender, slightly moveable mass on left; similar mass on right. *Diagnosis.* — Chronic inflammation of both tubes, with small ovarian cyst	Both tubes enlarged, occluded, and very firmly adherent. Right ovary normal, adherent; left ovary cystic, size of hen's egg, one cyst suppurating. Both tubes and left ovary removed	24 hours	Yes	R.	Some suppuration after removal of stitches at lower angle of wound. After 12th day recovery rapid. Sept. 2.—Stout and well. Oct. 22, 1892.—Well and strong; no pain; menstruates regularly; tendency to hernia in 2 or 3 places along wound.

VALUE OF ABDOMINAL SECTION IN

No.	Name.	Occupation and residence.	Age and civil condition.	Date of operation.	Place of operation.	Symptoms.	Duration of illness.
27	E. G.	Bookfolder, Lambeth	23 W.	1890 June 6	St. Thomas's Hospital	Severe pain lower part of abdomen, shooting down thighs, commencing 6 weeks after confinement. Symptoms subsided under hospital treatment in Dec., 1888. Well up to April, 1890, when she had to give up work owing to pain and hæmorrhage	18 months
28	E. L.	None, Streatham	34 M.	July 3	,,	Recurrent pelvic peritonitis; continuous pain for past 3 weeks in right iliac region and back	2 years
29	S. P	None, Stowmarket	27 S.	July 10	Private Nursing Home	Recurrent pelvic peritonitis, more frequent last 2 years. Dragging pain in right iliac region after least exertion. Loss of weight. Has been chiefly confined to bed past 5 or 6 weeks	7 years
30	A. T.	None, Peckham	24 M.	July 21	St. Thomas's Hospital	On July 5th severe attack of pain in left side, extending down leg; temp. 98·8° to 102·2°	13 days

CERTAIN CASES OF PELVIC PERITONITIS.   . 165

Physical signs and diagnosis.	Condition found and nature of operation.	Glass drainage-tube.	Peritoneum flushed.	Result.	Remarks.
Irregular, hard swelling in each posterior quarter of pelvis, more marked on right, where the tube can be felt thickened and the ovary prolapsed; swelling and tenderness in Douglas's pouch	Right tube thickened and adherent; right ovary enlarged, cystic, and adherent; left tube and ovary adherent; tube size of goose-quill. Under microscope, pus in contents of tubes. Both tubes and both ovaries separated and removed	46 hours	Yes	R.	Left hospital well in a month. April 2, 1891. —Quite well and free from pain. Has menstruated regularly last 6 months. Jan. 7, 1893. — Well and strong; married again 2 years ago; menstruates regularly; no pain except at menstrual period.
Tense, fluctuating swelling to left of uterus; on right, high up, a hard irregular swelling, giving the impression of tube and ovary embedded in a mass of adhesions	Pelvic contents matted together; on right side thickened tube and normal ovary densely adherent. During separation blood-stained pus escaped from amongst the adhesions. A thickened and prolapsed loop of large intestine adherent on left of uterus. Left tube and ovary not found. Right appendages removed; pus in right tube	44 hours	Yes	R.	Highest temp. after operation 100·4°. Left hospital well in a month.
Uterus fixed; irregular and hard mass on right side; less defined mass on left. *Diagnosis.*—Tubal disease with pelvic peritonitis	Pelvic contents densely matted. On right side mass of caseous material, partly inside and partly outside the tube, the tube having entirely lost its rugæ, and become separated by a ring of ulceration into two parts. Left side of pelvis also contained caseous material. Tube extremely adherent, occluded, thickened, and elongated. Both ovaries and both tubes removed	44 hours	No	R.	Did well first month, then had rise of temp., and eventually an abscess burst into rectum. Jan., 1891. — Very well; has lost almost all the aching pain after exertion. Nov. 17, 1892.—Feels very well; no pain since Spring.
Dense mass in each posterior quarter of pelvis, passing out from uterine cornu, and terminating as a thickened tube behind uterus. *Diagnosis.*—Double tubal disease, probably purulent, with peritonitis	Both tubes full of pus, and deeply ulcerated and perforated; walls very thick, distal ends closed; ovaries normal, adherent. Both tubes and ovaries removed	20 hours	Yes	R.	No pain on leaving hospital. Has gained flesh, and is in good spirits.

No.	Name.	Occupation and residence.	Age and civil condition.	Date of operation.	Place of operation.	Symptoms.	Duration of illness.
31	M. M.	None, Scarborough	23 M.	1890 Aug. 4	St. Thomas's Hospital	Admitted Feb. 1, 1890, with large pelvic hæmatocele, which disappeared. Returned to Scarborough March 25th. On Aug. 2 readmitted, not having been able to do much work on account of backache and pain in left iliac region	11 months
32	E. B.	None, Kent Road	40 M.	Aug. 5	,,	Continuous pelvic pain and dysmenorrhœa	6 months
33	Mrs. F.	None, Manchester	31 M.	Sept. 1	Private Nursing Home	Dyspareunia for several years. Attacked suddenly in July, 1890, whilst sitting reading out of doors, with extremely acute pelvic pain. In bed for a week, when severe symptoms recurred, followed by prostration, backache, flatulence, high temperature, and rapid pulse	4 years; acute symptoms 5 weeks
34	E. B.	None, Lambeth	29 M.	Sept. 4	St. Thomas's Hospital	Missed two menstrual periods; at third seized with aching pain in lower part of abdomen and back; face pale, features drawn; has been in bed since, and has had continuous slight hæmorrhage; pain has gradually diminished	6 weeks

Physical signs and diagnosis.	Condition found and nature of operation.	Glass drainage-tube.	Peritoneum flushed.	Result.	Remarks.
Irregular, hard mass in left posterior quarter of pelvis. *Diagnosis.*—Tubal disease	Both tubes and both ovaries matted by adhesions; tubes not enlarged, and but little thickened; remains of blood-clot behind uterus. Cavity cleansed, right ovary removed; right tube separated. Left tube and ovary not interfered with	24 hours	No	R.	Discharged Aug. 23rd; no pain; general condition improved. July, 1892.— Is in better health than for years. No pain or backache. Menses regular. Is quite fit for work.
Fixed, irregular swelling behind and beneath body of retroflexed uterus. *Diagnosis.* — Adherent tube and ovary behind retroflexed and adherent uterus	Prolapsed right ovary behind body of retroflexed uterus, enlarged from fibroids and incarcerated. No adhesions. Tubes and ovaries healthy. Displacement of uterus and ovary rectified. Pessary introduced *per vaginam*. Nothing removed	None	No	R.	August 29th.— Went home well; uterus in good position. April 18th, 1891. — Stout, well, and free from discomfort.
Left side of pelvis occupied by a fluctuating swelling rising into abdomen, and reaching to within 2¼ in. of umbilicus. Some prominence of abdominal wall above pubes. Uterus in front and to right fixed. *Diagnosis.*—Suppurating cyst of ovary and pelvic peritonitis	Pelvis occupied by matted viscera, with covering of omentum. Both Fallopian tubes enlarged and thickened; left tube stretched out over thick-walled suppurating cyst of left ovary, with which the tube was in direct communication at its fimbriated extremity. Contents of tube and ovary fetid. Right tube occluded. Right ovary indurated and slightly enlarged. Both tubes and both ovaries removed	36 hours	No	D.	Alarming amount of shock at close of operation. Died at 11.40 a.m., Sept. 5th. No P.M.
Firm tumour in left iliac region; uterus pushed upwards and forwards; length of uterine canal 3 in.; soft, irregular swelling behind uterus; thickening of right broad ligament	Ruptured blood-cyst of right broad ligament; intra-peritoneal hæmatocele; left tube distended with blood-clot; no trace of fœtus discovered. Cyst of broad ligament removed with right tube and ovary. Left tube removed	48 hours; then india-rubber tube 3 days	Yes	R.	Oct. 4th.—Discharged, looking and feeling well. Very slight discharge from sinus at lower angle of wound. Jan. 5th, 1892.— Well and at work ever since leaving hospital. Oct., 1892.—Stout and well; no pain; menstruates regularly.

VALUE OF ABDOMINAL SECTION IN

Name.	Occupation and residence.	Age and civil condition.	Date of operation.	Place of operation.	Symptoms.
L. B.	Mangler, Walworth Road	51 M.	1890 Sept. 9	St. Thomas's Hospital	Recurrent pelvic inflammation; pain in left side and yellow discharge
E. B.	Barmaid, Chelsea	34 S.	Oct. 16	,,	Chronic ill-health for years; severe pain and hæmorrhage 3 weeks ago after getting wet. Now complains of pain in left side of pelvis, shooting down thigh, and of slight hæmorrhage. Emaciated, very pale, and extremely ill
J. H.	None, Streatham	46 S.	Oct. 23	,,	Peritonitis after getting wet in Aug., 1889. Since then pain in pelvis, especially in right side, and after walking, standing, &c. Metrorrhagia. Symptoms worse last 3 months
M. N.	None, Battersea	51 M.	Nov. 10	,,	Profuse and irregular menstruation accompanied with pain, dating from puerperal illness 30 years ago. Great and continuous pain in back, especially on stooping and before defecation

Physical signs and diagnosis.	Condition found and nature of operation.	Glass drainage-tube.	Peritoneum flushed.	Result.	Remarks.
Uterus enlarged; on left side of pelvis a smooth, tense, fixed, elastic swelling, size of small orange. In right posterior quarter of pelvis an irregular, ill-defined swelling. *Diagnosis.* — Ovarian cyst on left; inflamed tube on right	Small ovarian cyst on left removed, with adherent but otherwise normal tube. Enlarged, prolapsed, and adherent tube removed, with normal ovary, on the right	48 hours	No	R.	Oct. 8th.—Left hospital well. 1891, Feb. 28th. —Presented herself at the hospital, looking well and in good condition. Has had little or no pelvic pain since operation. Has not menstruated. March 17th.—Small ventral hernia.
Uterus fixed; fixed, irregular, hard mass filling up left posterior quarter of pelvis, terminating behind uterus. *Diagnosis.*— Distended and adherent left tube and adherent ovary	Uterus and appendages of both sides involved in a mass of old adhesions. Both tubes thickened, containing muco-purulent fluid; outer coat of both ovaries thickened. Adhesions separated, both tubes and both ovaries removed	44 hours	No	R.	Improved rapidly. Went to convalescent home Nov. 22nd, where she gained 4½ lbs. in weight. Mar. 10th, 1893. — Is in better health than she has been for years; complains of flushes and occasional headache. Has not menstruated.
Irregular, hard swelling high up in left posterior quarter of pelvis, adherent to uterus. *Diagnosis.*—Inflamed tube and ovary, adherent	Both tubes thickened and enlarged, with thick, purulent mucus in their canal. Right ovary cystic, and dense from chronic inflammation; contents of cyst purulent and fetid. Both tubes and right ovary removed	44 hours		R.	Dec. 10th.—Left hospital stout, well, and free from pain. Feb. 27th, 1891. — Looks stout and well; complains of a little pain on right side. Some swelling and tenderness to right of uterus. In June, 1891, quite well. Died in November from cancer of stomach.
Soft swelling in pelvis in front of and to right of uterus; uterus fixed and retroverted. *Diagnosis.* — Cyst in pelvis, with chronic pelvic peritonitis	A number of thin-walled cysts of right broad ligament. Uterus retroverted and adherent. Tubes and ovaries bound down by old adhesions. Cysts of broad ligament removed. Adherent appendages not disturbed	48 hours	Yes	D.	Continued vomiting and abdominal distension. Died in a state of collapse on the 21st Nov., having complained of intense pain for four hours previously, No P. M.

No.	Name.	Occupation and residence.	Age and civil condition.	Date of operation.	Place of operation.	Symptoms.	Duration of illness.
39	S. R.	Ironer, Battersea	33 S.	1890 Nov. 12	St. Thomas's Hospital	Pelvic peritonitis in Adelaide Ward, April, 1889, when she had a discharge of pus from rectum. Left hospital June 8th, and remained well for two months. Since then had constant desire to defecate, and passed pus	19 months
40	A. B.	Servant, Brixton	22 S., 1-para	Nov. 19	,,	Pain in left iliac region, dating from 9th day after confinement. Frequent hæmorrhages. For past 6 weeks pain severe, discharge of blood continuous and profuse	6 months
41	E. C.	None, Lambeth	25 M.	Nov. 28	,,	First admitted Dec. 2, 1889, 7 weeks after confinement, with history that a few hours previously had been seized with severe abdominal pain, faintness, and vomiting. Temp.100·6° to 102·6°. Urine $\frac{1}{4}$ to $\frac{1}{10}$ albumen. Discharged much better Jan. 22, 1890. Re-admitted Nov. 12 with recurrence	12 months

Physical signs and diagnosis.	Condition found and nature of operation.	Glass drainage-tube.	Peritoneum flushed.	Result.	Remarks.
Hard and smooth swelling behind and to right of uterus. Evacuations contain pus. Diagnosis.—Suppurating ovarian cyst communicating with rectum	Small, inflamed, thick-walled, tense, and firmly adherent suppurating cyst of right ovary. Right tube inflamed. Left tube and ovary adherent, otherwise healthy. Right tube and suppurating cyst removed	48 hours; replaced by india-rubber tube	Yes	R.	Suppuration from wound for 7 weeks. Temperature after operation only once exceeded 99·4°; it was 100° on Nov. 21st from bowel disturbance. Discharged well Dec. 31st. April 18th, 1891. —No pain in pelvis or discharge from the bowel since leaving hospital; menstruates regularly.
Purulent discharge from cervix; cervical erosion; uterus retroverted; anterior to and below body of uterus, on left side, a well-defined oblong mass depressing left fornix, and divided into two portions by a sulcus. Right side free. Diagnosis.—Diseased left tube with normal ovary adherent	Thickened and unequally dilated left tube, containing thin pus, adherent; with the normal ovary to broad ligament and other parts. Uterus retroverted and adherent. Right tube normal; right ovary normal, but prolapsed and adherent. Left appendages removed. Uterus and right ovary set free	30 hours	No	R.	Discharged well Dec. 10, 1890. Readmitted in January, 1891, on account of some pelvic pain. Examined per vaginam, Jan. 16 and 28, with negative result. The temperature was normal. Evidently an instance of malingering.
Ill-defined soft mass behind and to right of uterus; smaller, harder, and irregular mass to left. Diagnosis.—Inflamed and adherent Fallopian tubes with diseased and enlarged right ovary	Pelvic viscera matted; right tube thickened and occluded; walls ¼ in. thick; no ulceration; no contents; mesosalpinx thickened; right ovary enlarged (2½ in. × 1¾ in.), on section found to be riddled with small abscesses. left appendages normal; right only removed; coil of intestine thickened and adherent in Douglas's pouch, during separation of which a small rent was made in the bowel; this was closed by sutures	24 hours	No	R.	Discharged Jan. 24, 1891, having gained flesh and with a good appetite; a slight purulent discharge from lower angle of wound. Only once (Dec. 5) was there a fæcal stain on dressing. Feb. 17.—Sinus not quite healed; menstruated for first time Feb. 13 to 16.

No.	Name.	Occupation and residence.	Age and civil condition.	Date of operation.	Place of operation.	Symptoms.	Duration illness.
42	A. H.	Mother's help, Brixton	20 S.	1890 Nov. 22	St. Thomas's Hospital	Continuous hæmorrhage for 2 months, commenced suddenly with a profuse flow, 2 weeks after a period, as she was carrying coals. Occasionally a little pain at lower part of abdomen. No loss of flesh; no pallor; no interference with general health	2 months.
43	E. S.	None, Richmond	25 M.	Dec. 17	,,	Pain in lower part of abdomen, back, and thighs, especially after standing. Gradual loss of strength and flesh. For 8 months unable to do housework; for last 4 months has been obliged to lie down almost entirely. Temp. normal	26 month
44	E. J. S.	None, Battersea	25 M.	Dec. 18	,,	Abdominal pain and weakness; loss of flesh; thick yellow vaginal discharge; pain on micturition. Has had to lie up frequently. (Husband in Clayton Ward in August, 1890, for urethral stricture)	2 year

Physical signs and diagnosis	Condition found and nature of operation	Glass drainage-tube	Peritoneum flushed	Result	Remarks
...is normal; body left, neck to right. right of uterus, on ...ane slightly pos- ...or to it, a soft, defined swelling. ...gnosis. — Hæma- ...a of broad liga- ...rt. (The swelling observed gradu- ...l; to increase in ...; operation a ...th after admis- ...)	Abscess in sheath of right rectus, 1¼ fl. oz., thick curdy pus evacu- ated. Parietal and visceral peritoneum everywhere studded with miliary tuber- cles. Large, soft, fluctuating sessile mass lying deeply in each posterior quarter of pelvis. Structures implicated not differ- entiated. Abdominal incision closed	25 hours	No	R.	Readmitted March 9 with emaciation and hectic; no change in physical signs. A year after operation in good health; no physical signs of dis- case now detected anywhere. Oct. 22, 1892. Is again losing flesh and feeling weak. No definite signs of disease either in abdo- men or pelvis.
...is displaced to ...; large, tender, ...lated swelling be- ...and to right; both ...ices (lateral) de- ...sed. Sulcus be- ...n right lateral posterior portions swelling. Dia- ...is.—Double pyo- ...nx	Both tubes enormously enlarged, occluded, and distended with thick pus; circum- ference of right 4½ in., of left 6½ in.; mu- cous membrane ulcer- ated. Both tubes se- parated and removed. Ovaries not seen	52 hours, replaced by India- rubber tube 20 hours	Yes	R.	Prolonged and severe shock after operation. Highest temp. during convalescence 100·2°. Discharged Jan. 24, 1891. March 6, 1891. —Looking and feeling well; has menstruated twice; no pelvic pain. Gonorrhœal vaginitis. Sept. 15.—Had influ- enza in May, not well since; nothing abnor- mal in pelvis; men- struation regular.
...tted August 11, ...l. Both tubes felt ...kened and adhe- ...nt. Improved great- ...y ...hospital. Went ...n August 30. Re- ...tted Dec. 15, ...ng been laid up ...discharge. Even, ...tender swelling ...nd and to right of ...us. Thickened along free border ...t broad ligament	Right tube enlarged and adherent, circum- ference 5¾ in., filled with old adherent clot, which protruded from open fimbriated end; outside tube a quantity of dark firm clot. Enlarged veins, filled with clot, seen beneath mucous lining of tube. Left tube occluded, otherwise normal. Both tubes and both ovaries re- moved	60 hours	Yes	R.	Suppuration in pelvis during convales- cence, pus discharged through lower angle of wound. Left hos- pital Jan. 18, 1891; very little discharge. Feb. 17.—States that sinus closed on Feb. 7; quite well; nothing abnormal per vagi- nam.

VALUE OF ABDOMINAL SECTION IN

Name.	Occupation and residence.	Age and civil condition.	Date of operation.	Place of operation.	Symptoms.
I. E.	None, Bermondsey	32 M.	1891 Jan. 8	St. Thomas's Hospital	Yellow discharge; bearing-down pain; pain in left iliac region, chiefly on standing or walking. Dysmenorrhœa; irregular menstruation; pain on micturition. Seriously ill for one week; acute pain on left side and diarrhœa
K. W.	Machinist, Peckham	23 S.	Jan. 15	,,	Attacked suddenly with "forcing pains" in abdomen. Two months later got her feet wet, and was seized with crampy pains in lower part of abdomen. Has been in bed a fortnight. Temp., day of admission, 100·6° to 104·2°
C. P.	Charwoman, Peckham	31 M.	Jan. 22	,,	Two months after confinement seized suddenly with severe pain in left iliac region and down thigh. More or less subject to similar attacks ever since. Last month much worse, with loss of flesh; pain on defecation. No disturbance of menstruation

Physical signs and diagnosis.	Condition found and nature of operation.	Glass drainage-tube.	Peritoneum flushed.	Result.	Remarks.
Uterus retroverted and displaced to right. To left and posteriorly a mass separated by a sulcus from body of uterus. On right some thickening ¼ in. in breadth, feeling like a coil of intestine. *Diagnosis.*—Double tubal disease, probably purulent	Pelvic viscera matted. Left tube elongated, thickened, twisted, and full of pus. Right tube thin-walled and tense, being distended with serum. Ovaries adherent but normal. both tubes and both ovaries removed	24 hours	No	R.	Recovery rapid. Discharged well Jan. 31. A month later had some pain on left side. Feb. 24.—A small swelling, size of normal ovary, on left side of pelvis; complains of pain on that side. On July 15 patient went into King's College Hospital, where, a few days later, Dr. Hayes removed a cyst of the left broad ligament. Nov. 13.—In Guy's Hospital, complaining of pain and desiring another operation; no discoverable lesion.
Uterus slightly displaced to right by a tender mass in left posterior quarter of pelvis size of small apple, depressing vaginal fornix	Left tube thickened and adherent, embracing enlarged ovary, size of pigeon's egg, containing a cyst full of blood. On section, wall of tube found three times its normal thickness; mucous membrane normal. Right tube and ovary normal. Left tube and ovary removed	No	No	R.	Rapid recovery. Discharged well Feb. 7. Mar. 26, 1892.—Attended on account of having lost flesh. No pain; menstruation regular. Uterus moveable; no abnormal swelling in pelvis.
Large mass high up in left posterior quarter of pelvis, traced from uterine cornu to back of cervix. Indistinct thickening in right posterior quarter of pelvis. *Diagnosis.*—Double tubal disease	Both tubes dilated and adherent. Conical blood-clot expanding outer inch of right tube, and continuous with small hæmatocele amongst the peritoneal adhesions. Hydrosalpinx of left tube	48 hours, replaced by rubber tube	Yes	R.	Discharged well Feb. 25.

No.	Name.	Occupation and residence.	Age and civil condition.	Date of operation.	Place of operation.	Symptoms.	Duration of illness.
48	K. W.	Still-room maid, Streatham	22 M.	1891 Jan. 29	St. Thomas's Hospital	Pain in right iliac region since birth of child 4 years ago. Eight months ago had a sudden escape of pus from vagina, which has continued to flow ever since	4 years
49	F. C. B.	Chambermaid, City	24 S.	Feb. 5	,,	Violated by stranger, 4th Nov., 1890. Five weeks later was admitted under Dr. Payne for pains in joints and fever. At end of January purulent vaginal discharge noticed, and patient complained of pelvic pain. Temp. 99° to 104·4°	7 weeks
50	M. W.	Servant, Wandsworth Road	19 S.	Feb. 26	,,	On Jan. 3 got wet through, and during night seized with severe pain in right iliac region. Was thought to have typhoid, and admitted to medical wards on Feb. 10. Constant headache, diarrhœa, hæmorrhage from bowel, no sickness	7 weeks

CERTAIN CASES OF PELVIC PERITONITIS. • 177

Physical signs and diagnosis.	Condition found and nature of operation.	Glass drainage-tube.	Peritoneum flushed	Result.	Remarks.
Douglas's pouch occupied by large, hard mass, extending more to right than left. Ring of dense hardness around cervix. Small aperture high up on posterior vaginal wall. *Diagnosis.*—Pelvic abscess with fistulous opening into vagina	Right ovary 2¼ in. × 1¾ in. × 1 in., consisting on section of a number of inflamed cysts, many of them full of pus. An opening, surrounded by granulation tissue, on surface, communicating with one of the abscess cavities. Whole mass adherent behind and below uterus. Right tube thickened. Left tube beaded from kinking, otherwise healthy. Right tube and ovary and left tube removed	20 hours	Yes	R.	Discharged well March 7. Nov. 15, 1891.—Readmitted on account of paroxysmal attacks of pain in right groin with vomiting. Between the attacks patient well and strong. Menstruates regularly. No abnormal swelling in pelvis; a little tenderness on right side. Temp. normal.
Purulent urethritis and purulent discharge from vagina; elongated tube-like swelling in right posterior quarter of pelvis; swelling less marked on left. Douglas's pouch occupied by cystic swelling. *Diagnosis.* — Gonorrhœal salpingitis	Pelvic viscera matted. Collection of pus in Douglas's pouch. Both tubes thickened, with pus in their canal, trickling from open fimbriated end into the retro-uterine abscess. Uterine appendages both sides removed. Appendix vermiformis removed	48 hours	Yes	R.	Temp. after operation 96°, 3 hours later 101·8°, at midnight 98·4°, after which never reached 100°. Joint pains disappeared, and patient quickly recovered her usual health.
On Feb. 19 irregular but somewhat elongated and smooth swelling in left posterior quarter of pelvis; less marked swelling on right. *Diagnosis.*—Purulent salpingitis on left with occlusion; on right, without. Gonorrhœal or tubercular	Thickened tube on left containing pus, and communicating by a recently ulcerated opening with the interior of a small suppurating cyst of adjacent ovary. Right tube enlarged and adherent. Right ovary normal. Both tubes and both ovaries removed	44 hours	Yes	R.	Recovery rapid.

[A report of the discussion on this paper appears in the 'Transactions of the Obstetrical Society,' vol. xxxiv, Fasc. 4 (Oct. to Dec., 1892). Abstracts were published in the 'Lancet' and 'British Medical Journal' for October 15th and November 19th, 1892. It was the Author's intention to reprint the discussion here, but some of the principal speakers having objected to further publicity being given to their remarks the intention has, in deference to their wishes, been abandoned.]

www.ingramcontent.com/pod-product-compliance
Lightning Source LLC
Chambersburg PA
CBHW020251170426
43202CB00008B/327